T0073111

MATHEMATICAL FOUNDATIONS OF INFORMATION SCIENCES

MATHEMATICAL FOUNDATIONS OF INFORMATION SCIENCES

Esfandiar Haghverdi

Indiana University Bloomington, USA

Liugen Zhu

Indiana University Indianapolis, USA

 World Scientific

NEW JERSEY · LONDON · SINGAPORE · BEIJING · SHANGHAI · HONG KONG · TAIPEI · CHENNAI · TOKYO

Published by

World Scientific Publishing Co. Pte. Ltd.
5 Toh Tuck Link, Singapore 596224
USA office: 27 Warren Street, Suite 401-402, Hackensack, NJ 07601
UK office: 57 Shelton Street, Covent Garden, London WC2H 9HE

Library of Congress Control Number: 2024002444

British Library Cataloguing-in-Publication Data
A catalogue record for this book is available from the British Library.

MATHEMATICAL FOUNDATIONS OF INFORMATION SCIENCES

ISBN 978-981-12-8888-3 (hardcover)
ISBN 978-981-12-9025-1 (paperback)
ISBN 978-981-12-8889-0 (ebook for institutions)
ISBN 978-981-12-8890-6 (ebook for individuals)

For any available supplementary material, please visit
https://www.worldscientific.com/worldscibooks/10.1142/13746#t=suppl

Desk Editors: Nandha Kumar/ Amanda Yun

Typeset by Stallion Press
Email: enquiries@stallionpress.com

To my wife Manijeh, my daughter Raya, and my son Iliad,
you are the brightest stars in my life.

E.H.

To my wife, Xiaofen Hu, for her unwavering support and love.

L.Z.

Preface

This book is based on the lecture notes originally prepared by the first author for the course I201-Mathematical Foundations for Informatics at Indiana University Bloomington. The first author taught from these notes for 10 years from 2002 until 2012. Starting in 2011 the second author adopted these notes in his teaching of the same course at Indiana University–Purdue University at Indianapolis (IUPUI). Since then, IUPUI faculty have created many new programs and I201 is now a required course for many of these programs, including BS in Informatics (online and in-person), Data Science, Applied Data & Information Sciences, Biomedical Informatics, and BA in Artificial Intelligence.

Informatics students come from a variety of backgrounds, for example, biology, social sciences, mathematics, and physical sciences, etc. It is not easy to design a course that fits the needs of everybody. These students will pursue different paths in their subsequent years, some will go to graduate school and some others might be joining the work force in industry. In designing this course we tried to understand the needs of such a diverse audience and decided to introduce topics that are vital for a successful completion of the informatics undergraduate program. The book you are reading now is the result of our efforts over the many years that we have been teaching this course.

The first two editions of this book were published by ClassPak Publishing, Indiana University. This new edition by World Scientific Publishing Co. Pte. Ltd. offers many new features, including many more worked examples, exercises (brand new with this edition), and a whole new chapter on Graph Theory.

For many of you it will be the first time you will see the material presented here. In general, when faced with unfamiliar mathematical definitions and techniques you might panic, this is entirely normal, however remember that these are the subjects you need to know to complete your studies successfully and more importantly perhaps to be equipped with tools of abstract thinking and reasoning that you shall find invaluable in life and work. Thus, what you need to do to overcome the initial panic and fear is to attack as many problems on each and every topic as you possibly can. Even if you cannot complete the solution to all problems, you will very often find that the mistakes you make, or the places you get stuck teach you a lot about the topic. Remember that

The only way to learn mathematics is to do mathematics.

Therefore, try to practice the skills you have learned on as many problems as you can. We have added many exercises to the end of each section in each chapter. There are also many solved problems. We recommend that you try to solve these worked exercises yourselves and refer to the solutions only when you are stuck and unable to make progress. We believe that the best way to interact with this book is to read it from the beginning to the end or to a point where you think your needs are satisfied, in sequential order. We do not assume any math skills beyond those learned in high school.

Despite our best efforts to remove all typos and errors, there are practically no books free of such. We would very much appreciate if our readers got in touch with us pointing out any such issues and more generally if they shared their experience with the book in any capacity. Please contact any one of us with your comments. We will maintain a support page for the book, including a list of errata at https://cgi.luddy.indiana.edu/~ehaghver/logic-book.

Acknowledgments

The first author has had many conversations with Professor Philip J. Scott from Department of Mathematics, University of Ottawa, Canada on topics related to these lectures. It is a pleasure to acknowledge the impact of his ideas, insight, and experience on the preparation of this book. We would also like to acknowledge the use of Paul Taylor's `diagrams.tex` to draw some diagrams and Peter Selinger's `fitch.sty` to typeset natural deduction proofs. It is a pleasure to thank Christopher B. Davis, Executive Editor, World Scientific Publishing for his constant encouragement and support during our work on this project.

Contents

Chapter 1

Introduction

In this book, we will study mathematical logic and some related topics that are typically classified as part of discrete mathematics. The primary objective of studying this material will be to acquire the art and science of

- analytical thinking,
- critical thinking,
- abstract thinking, and
- understanding and making *valid* arguments.

To this end, we begin by studying the simplest formal system that will help us formalize a text that involves reasoning in any given discipline. This system is called the *propositional logic* or *propositional calculus*. We will then study the shortcomings of this formal system and extend it to the richer language of *first-order logic*. This will be the most complicated formal system that we shall be studying in this book. Despite it not being adequate for informatics purposes, this shall provide enough background for applications that require more complicated formal systems that might be of use in more specialized fields in informatics or information sciences.

We will continue with the application of the techniques we learn to problems in mathematics, and then we shall study a powerful proof method in mathematics called *mathematical induction*. This will be followed by a study of important and very useful mathematical notions of functions and relations that have found applications in bioinformatics, chemical informatics, drug design, database theory, data-mining, etc. At the end, we touch upon graph theory where we follow an approach based on relations.

Here is a brief description of the contents of each chapter. In Chapter 2, we begin our study of the simplest formal system there is, that is propositional logic. We shall study truth tables, truth trees, and some logical puzzles. We will define and analyze validity of arguments in propositional logic. Next we describe a formal system that will allow us to generate proofs for valid arguments. Chapter 3 provides enough background on sets and set identities needed to introduce the more complicated formal system of predicate logic. We will carry an in-depth study of predicate logic, its syntax and semantics, translation from and to English, in Chapter 4. Chapter 5 will discuss the powerful mathematical proof technique, called mathematical induction. We will discuss this proof method in abstract and will provide many worked examples to make its use clear. In Chapter 6, we introduce and discuss the properties of functions and relations, these concepts play a central role in mathematics, computer science and all fields of informatics. It is vital that you get a good understanding and a working knowledge of these concepts. Finally, in Chapter 7, we give a brief discussion of graphs and some graph properties using a logical (relation-based) approach.

Chapter 2

Propositional Logic

In order to analyze the validity of arguments that we come across in science, mathematics, philosophy, politics, etc., we need to devise an artificial language, a formal system that will focus on some aspects of the language and ignore some other aspects. We shall also develop the machinery to model and analyze arguments in our formal language. The simplest formal language that one can imagine is *propositional logic*. It is powerful enough to capture propositions. However, it cannot talk about quantities, it cannot talk about time, etc. For these, we shall need to make our language stronger and richer. In this chapter, we will content ourselves with the simple formal system of propositional logic.

2.1 The Language of Propositional Logic

A *proposition* is a declarative sentence, that is to say, a sentence that is either true or false but cannot be both. Let's see some examples:

- Today is a beautiful sunny day.
- I am the King of France.
- 178917248747 is a prime number.
- All positive even integers greater than 2 can be expressed as the sum of two primes.

Note that even though establishing the truth or falsity of the propositions above may not be easy or even possible, we in principle know that they are either true or false. For example, the last proposition above is the famous *Goldbach conjecture*. We do not yet know if it is true or false.

On the other hand, the following are not propositions:

- $x + 2 = 4$.
- Do your homework!
- Is it still raining?
- This statement is false!

Let's see why: for the first statement above if we choose x to be 2, then we get $2 + 2 = 4$ which is clearly true, but if we choose x to be 3, then we get $3 + 2 = 5$ which is false. So, depending on the value of x, the statement $x + 2 = 4$ can be both false and true and hence is not a proposition. The second statement above is an imperative sentence and the third is a question, hence none is a proposition (a declarative statement). As for the fourth statement, we will let you think about it!

Let's recall that we are designing this artificial formal language to study arguments in natural languages. Hence, we must have ways of combining the propositions using connectives. The following table shows the five logical connectives. Here A and B are propositions.

Negation	$\neg$	$(\neg A)$	Not A
Conjunction	$\wedge$	$(A \wedge B)$	A and B
Disjunction	$\vee$	$(A \vee B)$	A or B
Implication	$\longrightarrow$	$(A \longrightarrow B)$	If A then B
Bi-implication	$\leftrightarrow$	$(A \leftrightarrow B)$	A if and only if B

Note that we include delimiting parenthesis so that more complicated proposition are easily and unambiguously readable. For example, applying the formation rules above we can form $((A \wedge B) \longrightarrow ((\neg A) \vee (\neg B)))$. If we remove the parenthesis, then we will get $A \wedge B \longrightarrow \neg A \vee \neg B$. Now it is not clear how we should read this formula! Of course, proceeding this way we will have complicated formulas crowded with many parenthesis and so logicians have devised precedence rules to disambiguate the reading of the formulas. Here is how it works: we say that connectives go from high to low precedence in the following list $\neg, (\longrightarrow, \leftrightarrow), (\vee, \wedge)$. So, negation has the highest precedence and conjunction and disjunction have the lowest precedence. For connectives of the same precedence (for example, $\wedge$ and $\vee$) we go from left to right. Let us look at some examples below.

Example 2.1.1 How shall we read $\neg A \vee B \longrightarrow C$?

Solution: *Given the precedence rule above, it should be read as* $((\neg A) \vee (B \longrightarrow C))$.

Example 2.1.2 How shall we read $\neg A \wedge B \vee \neg B \longrightarrow C$?

Solution: *Given the precedence rule above, it should be read as* $(((\neg A) \wedge B) \vee ((\neg B) \longrightarrow C)))$.

Exercises

Exercise 2.1.1 Which of the following statements are propositions?

1. Goldbach's conjecture is false.
2. Riemann's hypothesis cannot be settled.
3. Republicans will win the next three presidential elections.
4. Do not assume anything about politics.
5. Disinformation wars will not prove effective.
6. Should I buy more bitcoins?

Exercise 2.1.2 Put the parentheses around the formulas below that agree with the precedence rule above.

- $A \longrightarrow B \leftrightarrow C \wedge A \vee C$.
- $A \vee B \wedge C \leftrightarrow \neg A \wedge C$.

2.2   Truth Tables

Now that we can make compound propositions from simpler ones using connectives, and given that propositions are either true or false, we need to know how to determine the truth value of a compound proposition.

In the following A and B are formulas (propositions). We will use t and f to denote true and false, respectively. Here are the truth tables that define negation and conjunction connectives.

A	$\neg A$
t	f
f	t

A	B	$A \wedge B$
t	t	t
t	f	f
f	t	f
f	f	f

The truth value definitions for these connectives are intuitively very clear. They are what one would expect them to be. For disjunction we have,

A	B	$A \vee B$
t	t	t
t	f	t
f	t	t
f	f	f

The only row that demands some explanation is the first one where both A and B are true, this might seem strange to us, because normally when we use "or" in everyday language we use it exclusively. For example, we say "I will either stay home or go to the movies". Surely we do not consider doing both at the same time a possibility. Yet the connective above allows for both A and B to be true at the same time. That is why this connective is called *inclusive or*. To capture what we just alluded to above one can use the so called *exclusive or*, denoted $A \oplus B$ and read as "A or B but not both". It is defined by the following table:

A	B	$A \oplus B$
t	t	f
t	f	t
f	t	t
f	f	f

Let us continue with the implication.

A	B	$A \longrightarrow B$
t	t	t
t	f	f
f	t	t
f	f	t

There are many points here that we should discuss. First, note that from a logic point of view, we do not care about any causal or otherwise relationship between the *antecedent* A and the *consequent* B. We could say, "If $2 + 2 = 5$, then I am the King of France", and it will be true because $2 + 2 = 5$ is false. Secondly, it seems a bit strange that when the antecedent is false the implication is true, regardless of the truth value of the consequent. The reason is that logicians define the truth value for the implication based on deduction, that is if you start with a true premise (antecedent) and conclude something false, then your reasoning must be flawed, hence $A = t$, $B = f$, gives $A \longrightarrow B = f$, in all other cases one can find a correct logical reasoning to connect the antecedent to the consequent.

Next we shall look at the truth table for bi-implication connective.

A	B	$A \leftrightarrow B$
t	t	t
t	f	f
f	t	f
f	f	t

The idea is that bi-implication detects identical truth values. Note that it is true when A and B have the same logical value (either both are true, or both are false) and it is false otherwise.

Note that we do not need to use all these logical connectives, as a matter of fact we could only use negation and conjunction, or negation and disjunction among many other possibilities. In order to understand how we could do this, we need to have a notion of logical sameness (equivalence) and show that any formula is logically the same as a formula that uses just, say negation and conjunction. We will come back to this point after we have defined the notion of logical equivalence.

Let's look at some examples.

Example 2.2.1 For each of the formulas below we construct the truth table.

1. $A \vee \neg A$

A	$\neg A$	$A \vee \neg A$
t	f	t
f	t	t

2. $A \wedge (\neg A \vee \neg A)$

A	$\neg A$	$\neg A \vee \neg A$	$A \wedge (\neg A \vee \neg A)$
t	f	f	f
f	t	t	f

3. $A \longrightarrow (B \longrightarrow A)$

A	B	$B \longrightarrow A$	$A \longrightarrow (B \longrightarrow A)$
t	t	t	t
t	f	t	t
f	t	f	t
f	f	t	t

4. $A \longrightarrow (A \longrightarrow B)$

A	B	$A \longrightarrow B$	$A \longrightarrow (A \longrightarrow B)$
t	t	t	t
t	f	f	f
f	t	t	t
f	f	t	t

5. $A \longrightarrow ((B \wedge C) \vee \neg A)$

A	B	C	$\neg A$	$B \wedge C$	$(B \wedge C) \vee \neg A$	$A \longrightarrow ((B \wedge C) \vee \neg A)$
t	t	t	f	t	t	t
t	t	f	f	f	f	f
t	f	t	f	f	f	f
t	f	f	f	f	f	f
f	t	t	t	t	t	t
f	t	f	t	f	t	t
f	f	t	t	f	t	t
f	f	f	t	f	t	t

As you might have noticed above, some formulas are always true regardless of the truth value of their variables and some are always false. Yet some others are sometimes true and sometimes false. We shall distinguish these classes of formulas with the following definition.

Definition 2.2.2 A formula A is said to be:

- A *tautology*, if for all truth value assignments to its variables, its truth value is t. Or equivalently, if all the rows under the column of A in its truth table are t.
- A *contradiction*, if all the rows under its column are f, that is it is always false.

- A *contingency*, if the rows under the column of A are not all t and are not all f.
- *Satisfiable*, if at least one row under the column of A has the value t.

Here are some examples: $A \vee \neg A$ is a tautology, $A \wedge \neg A$ is a contradiction, whereas $A \wedge B$ is a contingency, and $A \longrightarrow B$ is satisfiable. You can easily determine these facts by looking at the truth tables of these formulas.

A natural question arises as how to determine that a given formula is a tautology or if it is a contradiction, etc. The definition above contains the answer. Let's look at some examples to clarify the matter.

Example 2.2.3 Consider the formula $A \vee (A \longrightarrow B)$ and check if it is a tautology.

Solution: *We form the truth table for this formula as follows:*

A	B	$A \longrightarrow B$	$A \vee (A \longrightarrow B)$
t	t	t	t
t	f	f	t
f	t	t	t
f	f	t	t

As can be seen from the table, the formula above is always true, so it is a tautology.

Example 2.2.4 Consider the formula $A \longrightarrow (A \longrightarrow B)$ and check if it is a tautology.

Solution: *We form the truth table for this formula as follows:*

A	B	$A \longrightarrow B$	$A \longrightarrow (A \longrightarrow B)$
t	t	t	t
t	f	f	f
f	t	t	t
f	f	t	t

As can be seen from the table above this formula is not a tautology. We shall give a counterexample to show that it can be made false. The counterexample is $\boxed{A = t, \ B = f}$ which is obtained from the second row of the table above.

Therefore, in order to check if a formula is a tautology, we form the truth table. If all rows under the formula have the value t, then it is a tautology, else it is not a tautology and any valuation for propositional variables that makes the formula false, is a counterexample. Note that if you claim that a formula is **not** a tautology, you need to support your answer by a counterexample showing that you can make the formula false.

What about contradiction or satisfiability?

Example 2.2.5 Consider the formula $P \longrightarrow \neg P$. Is this a contradiction?

Solution: *Let us construct the truth table for this formula.*

P	$\neg P$	$P \longrightarrow \neg P$
t	f	f
f	t	t

As can be seen from the table above, $P \longrightarrow \neg P$ is not always false, hence it is not a contradiction. We can give a counterexample, which will show that it can be indeed true, just let $P = f$. Note also that checking if a formula is a contradiction is quite similar to checking for tautology, in fact you can convince yourselves that a formula A is a tautology if and only if $\neg A$ is a contradiction.

Example 2.2.6 Consider the formula $P \wedge (\neg P \longrightarrow Q)$. Is this formula satisfiable?

Solution: *Let us construct the truth table for this formula.*

P	Q	$\neg P$	$\neg P \longrightarrow Q$	$P \wedge (\neg P \longrightarrow Q)$
t	t	f	t	t
t	f	f	t	t
f	t	t	t	f
f	f	t	f	f

As can be seen from the table above, this formula is true on rows 1 and 2, so it is satisfiable. Indeed, all we need is just one valuation for P and Q that makes the formula true and we can, for example say: let $P = t$ and $Q = f$. Therefore, to show that a formula is satisfiable you need to give one valuation for its variables making the formula true. On the other hand, if you claim it is not satisfiable, then you *need to construct the truth table for the formula and show that all rows have the value f.*

Exercises

Exercise 2.2.1 Draw the truth table for $(A \vee B) \longrightarrow B$. Is this formula a tautology? Is it satisfiable?

Exercise 2.2.2 Draw the truth table for $A \longrightarrow (A \vee B)$. Is this formula a tautology? Is it satisfiable?

Exercise 2.2.3 Draw the truth table for $(A \wedge B) \to (A \vee B)$. Is the formula a tautology?

Exercise 2.2.4 Find a counterexample for $A \vee (C \wedge \neg B)$ to show it is not a tautology.

Exercise 2.2.5 Is $\neg(P \to (Q \to P))$ satisfiable?

2.3 Truth Trees

Note that for a formula with n variables, the truth table will have 2^n rows because there are only two possibilities for each variable and there are n of them and hence by the multiplication rule, we have 2^n possible cases. As you can see from the examples above, to check if a formula is a tautology we have to consider all 2^n rows of its truth table, where n is the number of propositional variables in the formula. This is because there is no guarantee that if some proportion of rows are true, then all of them will be true. Similarly, if we wish to check if a formula A is satisfiable using tables, we might have to consider all 2^n rows, because it might just be that all $2^n - 1$ rows are false and that the very last one is true. We shall introduce a short-cut method based on the notion of *Truth Trees*. This method works more efficiently than truth tables in many cases of concern to us, however it should be noted that in special cases it can be as inefficient as the truth table method. At this time no one knows if the satisfiability problem can be solved efficiently, that is with an algorithm that takes a time to solve the problem in all cases that is a polynomial function of n (the number of propositional variables in the given formula).

The first thing to learn about a tree is how to grow it by branching, so we look at the branching rules.

Branching rules:

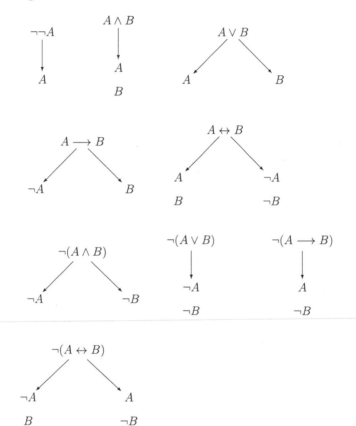

Before we start growing trees, let us talk about the motivation and the idea behind these rules. By inspecting these rules closely, one can observe that they have the following property:

> A formula at a node is true if and only if all the formulas under at least one of its branches are true.

For example, consider the formula $A \leftrightarrow B$, and suppose $A \leftrightarrow B$ is true then either both A and B are true, or they are both false. Conversely, if either both A and B are true, or both are false, then $A \leftrightarrow B$ will be true. This tells us how to form these rules, for example, if we wish to write the branching rule for $A \wedge B$, we will think of all the ways that we can make $A \wedge B$ true, but there is only one way and that is when both A and B are true, therefore in this case, we will have one branch only, with both A and B sitting underneath it. On the other hand, if we were to write the branching rule for $A \vee B$, we will put two branches with A underneath one of them and B underneath the other one. This is because to make $A \vee B$ true, we can either make A true or B true or both, but this latter case is included in either one of the previous cases, so we only have two branches.

Now that we have seen the rules let's grow some trees!

Example 2.3.1 Grow the tree of the formula $A \longrightarrow (A \longrightarrow B)$.

Solution:

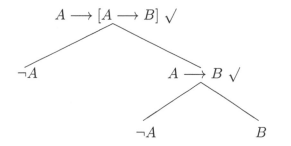

In this example, we have three leaves each with a single formula on them. No rule applies to the leaves, that is where we stop growing the tree. Hence, we continue growing the tree as long as there are formulas that can be broken down using the branching rules. Note also that after we break down a formula using branching rules, we put a check mark $\sqrt{}$ beside that formula to remind ourselves that we have already taken care of that node.

A *path* in a tree is a sequence of edges from the root to a leaf. A path is said to be *closed* if it contains a formula A and its negation $\neg A$. A path is said to be *open* if it is not closed.

The property of branching rules implies that every open path in a tree gives a way of making the formula at the root true. The way to do this is to assign the value t to each unchecked formula on a chosen open path. More generally, each open path in the truth tree of a formula A, corresponds to a row of the truth table of A where A is true.

Here is another example of a tree:

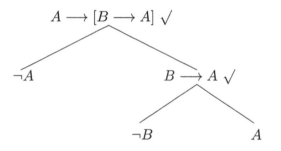

Here all paths are open. Note that one way to make the formula at the root true is to set $A = t$ with B being arbitrary, say it is $B = f$. This corresponds to the rightmost path. If we look at all the open paths, then we get all the possible ways to make the formula at the root true.

We shall next describe an algorithm to check a formula A for satisfiability.

2.3.1 *Satisfiability algorithm using truth trees*

Given a formula A, follow the following steps:

1. Put A at the root of the tree.
2. Apply an appropriate branching rule to an unchecked formula and check off this formula.
3. Close all the paths that you can by putting a cross ($\times$) underneath such a path. Do not grow the tree underneath a closed path.
4. Repeat steps 2–3 until one of the following happens:

a. All paths are closed.
b. There are open paths **and** all formulas (other than propositional variables or their negations) are checked

5. If 4(a) happens, then the formula A is NOT satisfiable. If 4(b)
   happens then the formula A IS satisfiable and every open path
   gives you an example that makes the formula A true.

Let us see how the algorithm works. Suppose we arrive in the case
where all paths are closed, then this means that there is no way at
all to make the formula at the root true, that is, the formula at the
root is always false and hence it is not satisfiable. On the other hand,
if there is at least one open path, that same open path gives us a
way of making the formula at the root true. Hence, this formula will
be satisfiable.

Here are some examples:

Example 2.3.2 Check if $\neg(A \longrightarrow (B \longrightarrow A))$ is satisfiable using
truth trees.

Solution: *We shall grow the tree*

$$\neg[A \longrightarrow [B \longrightarrow A]] \; \checkmark$$

$$A$$
$$\neg[B \longrightarrow A] \; \checkmark$$

$$B$$
$$\neg A$$
$$\times$$

*The single path here is closed due to A and $\neg A$ and since all paths
are closed then $\neg(A \longrightarrow (B \longrightarrow A))$ is NOT satisfiable.*

Example 2.3.3 Check if $A \longrightarrow (A \longrightarrow B)$ is a satisfiable using
truth trees.

Solution: *We shall grow the tree*

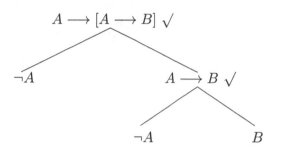

    All paths are open, so the formula $A \longrightarrow (A \longrightarrow B)$ is satisfiable. Note that, for example $A = t, B = t$ makes the formula true. Note that the fact that all paths are open has no bearing on our decision, all we care is for at least one path to be open.

    We next describe an algorithm that uses truth trees to check if a given formula A is a tautology. As a matter of fact checking A for tautology is the same as checking $\neg A$ for contradiction (showing that it is unsatisfiable). First, let's see why this is the case. Suppose A is a tautology, then it is always true and so $\neg A$ is always false, which means that $\neg A$ is unsatisfiable. Conversely, suppose $\neg A$ is unsatisfiable, then it is always false and hence A is always true and so it is a tautology. Thus, checking A for tautology is the same thing as checking $\neg A$ for satisfiability: if $\neg A$ is not satisfiable, then A is a tautology, and if $\neg A$ is satisfiable, then A is not a tautology. However, we rewrite the algorithm independently of this observation. Here it is.

2.3.2   *Tautology algorithm using truth trees*

Given the formula A, follow the following steps:

1. Put $\neg A$ at the root of the tree.
2. Apply an appropriate branching rule to an unchecked formula and check off this formula.
3. Close all the paths that you can by putting a cross ($\times$) underneath such a path. Do not grow the tree underneath a closed path.
4. Repeat steps 2–3 until one of the following happens:
   a. All paths are closed.
   b. There are open paths **and** all formulas (other than propositional variables and their negations) are checked.

5. If 4(a) happens then the formula A IS a tautology.
   If 4(b) happens then the formula A is NOT a tautology and every open path gives you a counterexample.

Let us look at some examples.

Example 2.3.4 Check if the formula $(A \longrightarrow B) \longrightarrow (A \wedge \neg B)$ is satisfiable using truth trees.

Solution: *Here is the tree for this formula:*

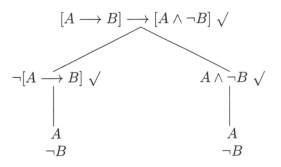

There is at least one open path and so the formula is satisfiable. The assignment $A = t$ and $B = f$ makes the formula true.

Example 2.3.5 Check if the formula $A \longrightarrow (B \longrightarrow A)$ is a tautology using truth trees.

Solution: *We shall grow the tree*

$$\neg[A \longrightarrow [B \longrightarrow A]] \checkmark$$
$$|$$
$$A$$
$$\neg[B \longrightarrow A] \checkmark$$
$$|$$
$$B$$
$$\neg A$$
$$\times$$

The single path here is closed due to A and $\neg A$ and since all paths are closed then $A \longrightarrow (B \longrightarrow A)$ is a tautology.

Example 2.3.6 Check if the formula $A \longrightarrow (A \longrightarrow B)$ is a tautology using truth trees.

Solution: *We shall grow the tree*

$\neg[A \longrightarrow [A \longrightarrow B]] \checkmark$

A
$\neg[A \longrightarrow B] \checkmark$

A
$\neg B$

There is an open path, so the formula $A \longrightarrow (A \longrightarrow B)$ is not a tautology and the counterexample can be found by setting all the variables along the open path to true and reading out the values. In this case we have $\boxed{A = t,\ B = f}$.

Example 2.3.7 Check if the formula $(A \lor B) \leftrightarrow (\neg A \longrightarrow B)$ is a tautology using truth trees.

Solution: *We shall grow the tree*

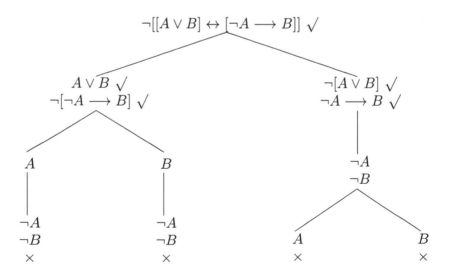

As you can see all paths are closed and so our formula is a tautology. In this example, we observe something new, namely that after the first branching rule we get two leaves each with two formulas on it. So, the natural question arises as to which one to break down first: the answer is that the order does not matter for our purposes. One more thing you should be careful about is that you have to break each formula underneath all the newly created branches that are descendants of that formula. That is to say on the left part of the tree we first break down $A \lor B$ and this way we create two new branches and then we break down $\neg(\neg A \longrightarrow B)$ under each one of them.

Exercises

Exercise 2.3.1 Draw the truth tree for $(A \lor B) \longrightarrow B$. Is this formula satisfiable?

Exercise 2.3.2 Draw the truth tree for $A \longrightarrow (A \land B)$. Is this formula satisfiable?

Exercise 2.3.3 Is $P \rightarrow (P \lor Q)$ a tautology? Please justify your answer with a truth tree or a counterexample.

Exercise 2.3.4 Is $P \lor (\neg P \rightarrow Q)$ a tautology? Please justify your answer with a truth tree or a counterexample.

Exercise 2.3.5 Is $(P \lor Q) \rightarrow (\neg P \land Q)$ a tautology? Please justify your answer with a truth tree or a counterexample.

Exercise 2.3.6 Is $(P \rightarrow Q) \lor (\neg P \rightarrow \neg Q)$ a tautology? Please justify your answer with a truth tree or a counterexample.

2.4   Logical Equivalence, and Consistency

Like in all other fields of mathematics, in logic we are also interested in defining a useful notion of sameness or equivalence. That is, because we do not want to deal with the same formula and redo all our analysis if it is the same as one we have already analyzed. Also, sometimes a formula which is complicated can have a very simple equivalent form that is easier to work with. There are many other applications of a good notion of sameness that we do not elaborate here. But the question is: what is a good definition of sameness? Let us look at an example, $A \wedge B$. This is certainly the same as $A \wedge B$ but that is not very useful because they are literally identical. Can we do better? Well, what about $B \wedge A$? Is this last one the same as the original formula $A \wedge B$? Some of you may say yes, and some may say no. For example, in daily use of natural language we may want to say A first because it has temporal significance, or we want to emphasize A over B, then the order of the letters matters. However, remember that in our artificial language of propositional logic we decided the only things we would care about were truth values. So, then $A \wedge B$ and $B \wedge A$ are the same because they are true and false precisely at the same time (check this!). This motivates the following definition.

Definition 2.4.1 (Logical equivalence) We say that formulas A and B are *logically equivalent*, denoted $A \equiv B$, iff the formula $A \leftrightarrow B$ is a **tautology**.

Note that this definition already gives us a method to check if two formulas A and B are equivalent, all we have to do is to check whether $A \leftrightarrow B$ is a tautology.

Let's do an example and show that $A \wedge B \equiv B \wedge A$. As can be seen from the table below the equivalence indeed holds:

A	B	$A \wedge B$	$B \wedge A$	$(A \wedge B) \leftrightarrow (B \wedge A)$
t	t	t	t	t
t	f	f	f	t
f	t	f	f	t
f	f	f	f	t

There are some important logical equivalences in logic with given names. We fix two constant formulas, T which is always true, and F which is always false. These can be any formulas that have the required property, for example, you can choose $T = A \vee \neg A$ and $F = \neg T$. The following table contains some very important logical equivalences (laws). We will use these to prove equivalences for more complicated formulas.

$A \wedge B \equiv B \wedge A$	Commutativity of $\wedge$
$A \vee B \equiv B \vee A$	Commutativity of $\vee$
$(A \wedge B) \wedge C \equiv A \wedge (B \wedge C)$	Associativity of $\wedge$
$(A \vee B) \vee C \equiv A \vee (B \vee C)$	Associativity of $\vee$
$A \wedge (B \vee C) \equiv (A \wedge B) \vee (A \wedge C)$	Distributivity of $\wedge$ over $\vee$
$A \vee (B \wedge C) \equiv (A \vee B) \wedge (A \vee C)$	Distributivity of $\vee$ over $\wedge$
$A \wedge A \equiv A$	Idempotence of $\wedge$
$A \vee A \equiv A$	Idempotence of $\vee$
$A \wedge T = A, \ A \vee F \equiv A$	Identity
$A \wedge F \equiv F, \ A \vee T \equiv T$	Domination
$\neg(A \wedge B) \equiv \neg A \vee \neg B$	De Morgan
$\neg(A \vee B) \equiv \neg A \wedge \neg B$	De Morgan
$A \longrightarrow B \equiv \neg A \vee B$	Implication
$A \leftrightarrow B \equiv (A \longrightarrow B) \wedge (B \longrightarrow A)$	Bi-implication
$A \vee \neg A \equiv T$	Tautology
$A \wedge \neg A \equiv F$	Contradiction
$\neg\neg A \equiv A$	Double negation
$A \vee (A \wedge B) \equiv A$	Absorption
$A \wedge (A \vee B) \equiv A$	Absorption

Let's check one of the logical equivalences above using truth trees.

Example 2.4.2 Check if $\neg(A \vee B) \equiv (\neg A \wedge \neg B)$.

Solution: *Clearly for this we need to check if $\neg(A \vee B) \leftrightarrow (\neg A \wedge \neg B)$ is a tautology. We shall grow the tree*

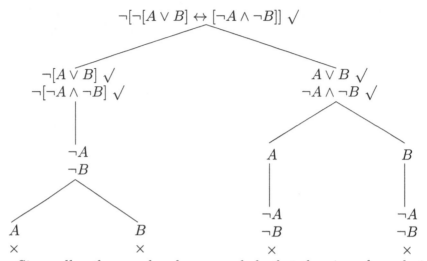

Since all paths are closed, we conclude that the given formula is a tautology.

Using the logical equivalences given in the table above, we can prove more complicated logical equivalences in an algebraic manner, as is demonstrated in the example below.

Example 2.4.3 Show that $P \longrightarrow (Q \longrightarrow R) \equiv Q \longrightarrow (P \longrightarrow R)$ using logical laws.

Solution:

$$P \longrightarrow (Q \longrightarrow R) \equiv \neg P \vee (Q \longrightarrow R), \textit{Implication} \qquad (2.1)$$

$$\equiv \neg P \vee (\neg Q \vee R), \textit{Implication} \qquad (2.2)$$

$$\equiv (\neg P \vee \neg Q) \vee R, \textit{Associativity} \qquad (2.3)$$

$$\equiv (\neg Q \vee \neg P) \vee R, \textit{Commutativity} \qquad (2.4)$$

$$\equiv \neg Q \vee (\neg P \vee R), \textit{Associativity} \qquad (2.5)$$

$$\equiv \neg Q \vee (P \longrightarrow R), \textit{Implication} \qquad (2.6)$$

$$\equiv Q \longrightarrow (P \longrightarrow R), \textit{Implication} \qquad (2.7)$$

Note that this also gives us a new and alternative way to prove that a formula is a tautology, namely given a formula A we prove it to be a tautology by showing that $A \equiv T$, where T is a logical formula that is always true. Here is an example.

Example 2.4.4 Show that $P \longrightarrow (Q \longrightarrow P)$ is a tautology using logical laws.

Solution: *We shall try to show that* $P \longrightarrow (Q \longrightarrow P) \equiv T$ *using logical laws.*

$$P \longrightarrow (Q \longrightarrow P) \equiv P \longrightarrow (\neg Q \vee P), Implication$$
$$\equiv \neg P \vee (\neg Q \vee P), Implication$$
$$\equiv \neg P \vee (P \vee \neg Q), Commutativity$$
$$\equiv (\neg P \vee P) \vee \neg Q, Associativity$$
$$\equiv T \vee \neg Q, Tautology$$
$$\equiv T, Domination$$

Early on in this chapter, when we introduced logical connectives, we mentioned that we did not need all of them and that we could have chosen, for example, to use only $\{\neg, \wedge\}$. Here we shall clarify what this means. If we decide to use only negation and conjunction connectives, then we need to show that all other connectives are logically equivalent to formulas built from negation and conjunction. Here they are,

- $A \vee B \equiv \neg(\neg A \wedge \neg B)$.
- $A \longrightarrow B \equiv \neg(A \wedge \neg B)$.
- $A \leftrightarrow B \equiv \neg(A \wedge \neg B) \wedge \neg(B \wedge \neg A)$.

We leave it to the reader to check that the equivalences above hold. A set like $\{\neg, \wedge\}$ is called *functionally complete*, because of the property that the connectives in it are enough to express all logical functions. There are other such sets, for example $\{\neg, \vee\}$ is another functionally complete set.

Now we switch gears and look at a very interesting application of what we have learned so far. It is a very important problem, called *logical consistency*. The formal definition is given below but let us for the moment think about what this is. Suppose I wish to have certain things in life and I write down my wishes. Here is an example:

1. I will buy a new car.
2. I will buy a new car if and only if I have money.

3. I will have money if and only if I save money.
4. I am not a good money saver.

Now having said all of these, you may wonder if it is possible for all these statements (propositions) to be true at the same time. As it turns out, and we will see how to do this in a systematic way below, these wishes cannot all come true at the same time. Let us have a look: I am not a good money saver so, by (3) I will not have money, and then by (2) I will not buy a car. But then, this is the opposite of (1). So, as you can see we can not make all these wishes come true at the same time. Oh well, perhaps I have to reconsider my car wishes or develop a saving habit.

Things can be much more serious and more complicated to check. Imagine we are talking about the specifications for an operating system or a communication protocol, or a security protocol. We need to make sure that all the statements can be made true at least in one way before we embark on designing the system according to the given specification.

Here is the formal definition.

Definition 2.4.5 (Consistency) A set $\{A_1, A_2, \ldots, A_n\}$ of formulas is said to be *consistent* if all the A_i can be made true at once. In other words, iff the formula $A_1 \wedge A_2 \wedge \cdots \wedge A_n$ is **satisfiable**.

For example, the set $\{A, A \longrightarrow B, B \vee C\}$ is satisfiable, we can take $A = t, B = t, C = t$ and this assignment of values to variables makes all three formulas in the set, true. However, the set $\{A, A \longrightarrow B, \neg B\}$ is not consistent because the formula $A \wedge (A \longrightarrow B) \wedge \neg B$ is a contradiction, that is, it is not satisfiable.

Clearly, we can check consistency using truth tables. For example, you can construct the table for the formula $A \wedge (A \longrightarrow B) \wedge \neg B$ and see that it has value f on all rows.

However, this is a very long method and becomes very unmanageable with many variables. So, we shall use the method of trees.

2.4.1   *Algorithm to check the consistency of* $\{A_1, \ldots, A_n\}$ *using trees*

1. Put the formulas $A_1, A_2, \ldots, A_n$ underneath each other at the root of the tree.

2. Grow the tree.
3. If all paths are closed then the set is NOT consistent (that is, it
   is inconsistent). Else, that is, if there is at least one open path,
   then the set is consistent and assignment of t to all the variables
   on this open path will give you a way to make all the formulas at
   the root true.

Example 2.4.6 Show that the set $\{A, A \longrightarrow B, B \vee C\}$ is consistent.

Solution:

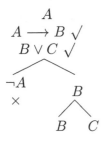

As you can see there is an open path. For example, setting
$A = t, B = t, C = t$ will make all three formulas true.

Example 2.4.7 Show that $\{A, A \longrightarrow B, \neg B\}$ is inconsistent.

Solution:

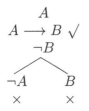

As you can see there are no open paths, so the set is not
consistent.

We continue with the definition of an argument and the notion
of validity in the next section. We will also discuss the methods to
check the validity of arguments.

Exercises

Exercise 2.4.1 Use a truth tree to show $A \wedge B \equiv B \wedge A$.

Exercise 2.4.2 Use a truth tree to show
$$P \longrightarrow (Q \longrightarrow R) \equiv Q \longrightarrow (P \longrightarrow R).$$
Exercise 2.4.3 Is the set $\{(A \vee B) \longrightarrow B, A, \neg B\}$ consistent? Either give an example (truth values for all the variables) to show that it is, or prove that it is not.

Exercise 2.4.4 Is the set $\{(A \wedge B) \longrightarrow B, A, \neg B\}$ consistent? Either give an example (truth values for all the variables) to show that it is, or prove that it is not.

Exercise 2.4.5 Show that the set $\{(A \longrightarrow B) \longrightarrow B, A \longrightarrow B, \neg B\}$ is inconsistent.

2.5 Arguments and Validity

In this section, we extend our reach and look at an even more exciting application, checking the validity of logical arguments. Again, let us start with an informal example and see what this is all about.

Let us make some reasoning and form a logical argument. We will do so in English. Suppose: 1. Today is a cold day; 2. If a day is a cold day, then it rains; 3. On rainy days I do not go out. Therefore, I will not go out today.

Note that first I said a few things (three things to be precise) and I want you to believe that if these three things I said were true, then what I said after "Therefore" must also be true. That is, if you believe everything I said before making my final statement (conclusion), then you have to believe my conclusion as well. But why? Well, if the logical argument I am making is valid, then you have to believe my conclusion out of logical necessity. That is to say, the truth of my conclusion is forced on you by laws of logic. What if my argument is not logically valid? Well, then there is no reason for you to believe in my conclusion, even though you agree with me about everything I said before that.

The argument above is actually valid (you will see how to check for that below after the formal definition). Note that you may disagree with the statement that it rains on cold days. But that is not the point, and is not relevant. My argument, if it is valid means that if you agree with everything I say, then you cannot possibly (logically) disagree with my conclusion.

Definition 2.5.1 An *argument* is a set of sentences, in which one sentence called the *conclusion* is claimed to follow from the other sentences called the *premises* or the *hypotheses*. A *logical argument* is the translation of an argument into the language of propositional logic. We shall use the following format to represent a logical argument:

$$A_1$$
$$A_2$$
$$\vdots$$
$$\frac{A_n}{C}$$

here the A_i are the hypotheses and C is the conclusion.

We say that a logical argument as above is *valid* iff the formula

$$(A_1 \wedge A_2 \wedge \cdots \wedge A_n) \longrightarrow C$$

is a tautology. Else, we say that the argument is *invalid*. Therefore in the case that an argument is invalid we will have assignments to variables such that all the A_i are true and yet the conclusion C is false. This constitutes a *counterexample* to the validity of the argument.

Note that checking the validity of an argument is by definition reduced to the problem of checking for tautology, something that we already know how to do. Nevertheless, we shall give a simple algorithm below which is slightly shorter.

2.5.1 *Algorithm for checking validity using truth trees*

1. Put $A_1, A_2, \ldots, A_n$ and $\neg C$ underneath each other at the root of the tree.
2. Grow the tree.
3. If all paths are closed, then the argument is valid. Else, that is if there is at least one open path, then the argument is invalid and any open path will give a counterexample.

Let's see how this algorithm works. Suppose all paths are closed then it means that the formula $\neg[(A_1 \wedge \cdots \wedge A_n) \longrightarrow C]$ cannot be made true, so it is always false, so the formula $(A_1 \wedge \cdots \wedge A_n) \longrightarrow C$ is a tautology. Similarly try to convince yourselves that the algorithm works properly when there is an open path.

Let's look at some examples.

Example 2.5.2 Check the validity of the argument: $\dfrac{\begin{array}{c} A \longrightarrow B \\ A \end{array}}{B}$.

Solution: *Using the method of truth tables we can easily see that $((A \longrightarrow B) \wedge A) \longrightarrow B$ is a tautology and hence the argument is valid.*

Example 2.5.3 Check the validity of the argument: $\dfrac{\begin{array}{c} A \longrightarrow B \\ \neg A \end{array}}{\neg B}$.

Solution: *Let's use truth trees to check if this argument is valid.*

There is an open path so the argument is invalid. Indeed if we let $A = f, B = t$ we see that all the hypotheses become true while the conclusion becomes false.

Example 2.5.4 Check the validity of the argument: $\dfrac{\begin{array}{c} A \longrightarrow B \\ B \end{array}}{A}$.

Solution: *Let's use truth trees to check if this argument is valid.*

There is an open path so the argument is invalid. Indeed if we let $A = f, B = t$ *we see that all the hypotheses become true while the conclusion becomes false.*

Example 2.5.5 Check the validity of the argument:

$$\begin{array}{c} A \\ A \longrightarrow B \\ B \longrightarrow C \\ \hline C \vee D \end{array} .$$

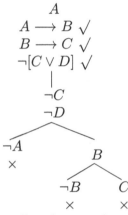

As all paths are closed, the argument is valid.

Example 2.5.6 Check the validity of the argument:

$$\begin{array}{c} A \\ A \longrightarrow B \\ B \longrightarrow C \\ \hline C \wedge D \end{array} .$$

Using truth trees:

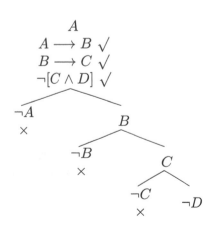

There is an open path, therefore the argument is invalid. A counterexample is $A = t, B = t, C = t, D = f$.

Exercises

Exercise 2.5.1 Show that the argument we made in the opening of this section is valid.

Exercise 2.5.2 Check the validity of the argument $\dfrac{\begin{array}{l}(A \vee B) \longrightarrow B \\ A\end{array}}{B}$.

Note that, you have to either prove that it is valid or give a counterexample showing that you can make the hypotheses true and the conclusion false.

Exercise 2.5.3 Check the validity of the argument $\dfrac{\begin{array}{l}(A \vee B) \longrightarrow B \\ C \longrightarrow A \\ C\end{array}}{\neg B}$.

Note that, you have to either prove that it is valid or give a counterexample showing that you can make the hypotheses true and the conclusion false.

Exercise 2.5.4 Show that the given argument is invalid: $\dfrac{\begin{array}{l}(A \leftrightarrow B) \\ C \wedge \neg A\end{array}}{B}$.

Exercise 2.5.5 Show that the given argument is valid: $\dfrac{\begin{array}{l}(A \leftrightarrow B) \\ C \wedge \neg A\end{array}}{\neg B}$.

2.6   Logical Puzzles

In this section, we will look at some logical puzzles. We deserve to have some fun; we have worked so hard on doing lots of cool things with our simple artificial language.

These puzzles are adaptations from those in the book by Raymond Smullyan entitled *What is the name of this book?* These puzzles are about the inhabitants of a Pacific island called the Island of Knights and Knaves. There are two groups of inhabitants on this island: Knights who always tell the truth and Knaves who always lie. We shall assume in all of the examples below that any inhabitant is either a knight or a knave. Therefore, if we can conclude that some inhabitant, say A is not a knave then she must be a knight.

Example 2.6.1 Suppose A says, "I am a knave but B isn't". What are the types of A and B?

Solution: *Suppose A is a knight then what she says is true, that is its truth value is t, but the proposition "I am a knave" is then false making the whole statement false, we get a contradiction. So, A is not a knight and hence she must be a knave. Then what she says has the value f, also the proposition "I am a knave" has the value t, so "B isn't a knave" has to be false meaning that B is a knave. Therefore, both A and B are knaves.*

Example 2.6.2 A says, "If I am a knight, then $2+2=4$". Can you determine the type of A?

Solution: *Note that $2 + 2 = 4$ is always true, so the value of the implication in the quotation mark is true and hence A cannot be a knave and thus he is a knight.*

Example 2.6.3 A reporter asks A, "Are you a knight?" and A says, "If I am a knight, then I'll eat my hat". Show that A has to eat his hat.

Solution: *We do this by proving that he has to be a knight. In other words we show that he cannot be a knave. Suppose he is a knave, then the statement "I am a knight" is false making the whole statement in quotation marks true, contradicting his being a knave. So, A is a knight and what he says is true and the statement "I am a knight" is true, so "I'll eat my hat" must be true. Thus, he has to eat his hat.*

Example 2.6.4 We have three people A, B and C. A says, "B and C are of the same type". Someone then asks C, "Are A and B of the same type?" What does C answer?

Solution: *Suppose A is a knight, then either B and C are both knights, or they are both knaves. If they are both knights, then as A and B are of the same type and C is a knight, so C will say YES. If both B and C are knaves then A and B are of different types, but C being a knave will again say YES.*

Now suppose A is a knave, then either B is a knight and C is a knave, or B is a knave and C is a knight. In the first case, A and B are of different types, but C being a knave will say YES. In the second case, A and B are of the same type and C is a knight, so C will again say YES. Thus, in all cases C replies YES to the question.

Exercises

Exercise 2.6.1 We have three inhabitants A, B, and C on the Island of Knights and Knaves. A and B make the following statements:

A: "I am a knave and B is a knight".
B: "Exactly one of the three of us is a knight".

Can you determine what A, B and C are?

Exercise 2.6.2 We are on the Island of Knights and Knaves again. In each case can you determine the type of A and B?

a. A says, "We are both knaves". And B says nothing.
b. A says, "I am a knave or B is a knight". And B says nothing.

2.7   Translation to Propositional Logic

We say that a sentence in English is truth-functionally atomic or just atomic, if there are no connectives used in the sentence (see below for examples.) The truth-functionally atomic sentences in English are translated as atomic propositions for which we use capital letters. Here are some examples of atomic sentences:

• This blackboard is white.
• The room temperature is too hot for us to work for long hours without getting tired very quickly.
• The glass is half full.
• There are two tables in this classroom.

Note that an atomic sentence in our sense (i.e., truth functionally atomic) may be quite complicated otherwise. The point is that one should not have any logical connectives in an atomic sentence.

The translation process for composite sentences is done in some steps:

1. Undorlino tho koy logical connoctivos, romombor that oach logical connective might have many stylistic variations in English. See Appendix A for stylistic variations of connectives.
2. Make sure that the parts that are not underlined are atomic sentences and associate letters with them.
3. Complete the translation by putting in the connectives.

Example 2.7.1 Translate the following sentence into prop. logic:

> *If $3^{13} + 1$ is an odd number, then either it is a prime or the product of two odd numbers.*

Solution: *The atomic sentences together with their translations are:*

- *O: $3^{13} + 1$ is an odd number.*
- *P: $3^{13} + 1$ is a prime number.*
- *N: $3^{13} + 1$ is the product of two odd numbers.*

And we get

$$O \longrightarrow (P \vee N)$$

as the translation of the sentence.

Example 2.7.2 Translate the following sentence into prop. logic:

> *If it snows or freezes tomorrow, then if the trees are in blossom and are unprotected, then the crop will be ruined unless a miracle occurs.*

Solution: *The atomic sentences together with their translations are:*

- *S: It snows tomorrow.*
- *F: It freezes tomorrow.*
- *B: The trees are in blossom.*
- *U: The trees are protected.*
- *C: The crop will be ruined.*
- *M: A miracle occurs.*

And we get the formula

$$(S \vee F) \longrightarrow ((B \wedge \neg U) \longrightarrow (C \vee M)).$$

Example 2.7.3 Translate the following sentence into prop. logic:

> *If Len gets sick, if he drinks too much, then given that he is healthy if and only if he is sober, he drinks too much, if he gets sick.*

Solution: *The atomic sentences together with their translations are:*

- *S: Len gets sick.*
- *D: Len drinks too much.*
- *H: Len is healthy.*
- *B: Len is sober.*

And we get the formula

$$(D \longrightarrow S) \longrightarrow ((H \leftrightarrow B) \longrightarrow (S \longrightarrow D)).$$

Example 2.7.4 Check the validity of the following argument.

> *Either it is not the case that Leslie pays attention and does not lose track of the argument, or it is not the case that she does not take notes and does not do well in the course. Leslie neither does well in the course nor loses track of the argument. If Leslie studies logic, then she does not do well in the course only if she does not take notes and pays attention. Therefore Leslie does not study logic.*

Use the following atomic propositions.

P: Leslie pays attention.
L: Leslie loses track of the argument.
N: Leslie takes notes.
W: Leslie does well in the course.
S: Leslie studies logic.

Solution: *The logical argument is*

$$\neg(P \wedge \neg L) \vee \neg(\neg N \wedge \neg W)$$
$$\neg(W \vee L)$$
$$\frac{S \longrightarrow (\neg W \longrightarrow (\neg N \wedge P))}{\neg S}.$$

Using the method of truth trees one can see that this is a valid argument. Here is another method to check the validity: suppose the *argument is invalid that is all the hypotheses are true and the conclusion is false, then we get that $S = t$ and that $W = f$ and $L = f$ and from the third hypothesis we get that $N = f$ and $P = t$, plugging in these values in the first hypothesis we get a contradiction. Therefore, the argument is valid.*

We continue with some more examples of arguments.

Example 2.7.5 Check the validity of the following argument.

> *John's girlfriend will be at the recital. If his girlfriend is at the recital, and John makes noise, she will be unable to hear the music, and so will be displeased. John's girlfriend will not be displeased. Therefore, John will keep quite at the recital.*

Use the following atomic propositions:

G: *John's girlfriend will be at the recital.*
N: *John makes noise.*
M: *John's girlfriend will hear the music.*
D: *John's girlfriend will be displeased.*

Solution: *The logical argument is*

$$G$$
$$(G \wedge N) \longrightarrow (\neg M \wedge D)$$
$$\frac{\neg D}{\neg N} \ .$$

Using the method of truth trees one can see that this is a valid argument. Using the method of contradiction as in the example above we get that $N = t$, $G = t$ and $D = f$ but then the second hypothesis will give a contradiction. Therefore, the argument is valid.

Example 2.7.6 Are the following statements consistent?

> *The router can send packets to the edge system only if it supports the new address space. For the router to support the new address space, it is necessary that the latest software release be installed. The router can send packets to the edge system if the latest software release is installed. The router does not support the new address system. (Use R, A, I)*

Solution: *We get the set* $\{R \longrightarrow A, A \longrightarrow I, I \longrightarrow R, \neg A\}$. *Using truth trees we see that the set is consistent, for example:* $R = f$, $A = f, I = f$ *makes all the formulas true.*

Example 2.7.7 Check the validity of the following argument.

> If I study law, then I will make a lot of money. If I study archae-
> ology, then I will travel a lot. If I make a lot of money or travel
> a lot, then I will not be disappointed. Therefore, if I am dis-
> appointed, then I neither studied law nor studied archaeology".
> *(Use* L, M, A, T, D)

Solution: *The logical argument is*

$$L \longrightarrow M$$
$$A \longrightarrow T$$
$$\underline{(M \vee T) \longrightarrow \neg D}$$
$$D \longrightarrow (\neg L \wedge \neg A)$$

Using the method of contradiction, we get that $D = t$ *and* $\neg L \wedge \neg A = f$ *and from the third hypothesis we see that* $M = f$ *and* $T = f$ *and then from the first hyp.* $L = f$ *and from the second hyp., we get* $A = f$. *Then* $\neg L \wedge \neg A = t$ *gives a contradiction, since we had this as false. So, the argument is valid.*

In some of the examples above we used a method to check the validity of the argument that we called the method of contradiction. We shall be a bit more explicit about this method. Suppose you are given an argument with hypotheses A_i and conclusion C and you wish to check this argument for validity. Here is how the *method of contradiction* works, you assume that the argument is invalid, that is all the hypotheses are true and the conclusion is false. If you succeed to realize this situation, then the argument is indeed invalid and you have actually found a counterexample to its validity. However, if you fail to realize this situation, that is you face a contradiction, then the argument is valid.

Exercises

Exercise 2.7.1 Translate the statement "I will get an A only if I attend every class" to propositional logic.

Use the following atomic propositions:

A: I will get an A,
C: I attend every class.

Exercise 2.7.2 Translate the statement "You will get admitted into your dream university only if your high school GPA is higher than 2.5", to propositional logic.

Use the following atomic propositions:

U: You will get admitted into your dream university,
G: Your high school GPA is higher than 2.5.

Exercise 2.7.3 Translate the following statements to propositional logic.

(i) It snows whenever the wind blows from the northeast.
(ii) For you to win the contest it is necessary and sufficient that you have the only winning ticket.
(iii) I will remember to send you the address only if you send me an e-mail message.
(iv) Grizzly bears have not been seen in the area and hiking on the trail is safe, but berries are ripe along the trail.

Exercise 2.7.4 Consider the following argument:

> *If Miranda does not take a course in logic, then she will not graduate. If Miranda does not graduate, then she is not qualified for the job. If Miranda reads this book, then she is qualified for the job. Therefore, Miranda does not take a course in logic but she reads this book.*

Is this argument valid? If yes, give a proof, if not, give a counterexample.

Use the following atomic propositions:

L: Miranda takes a course in logic,
G: Miranda will graduate,
Q: Miranda is qualified for the job,
R: Miranda reads this book.

Exercise 2.7.5 Translate the following statements to propositional logic and check if they are consistent.

> *Bob will study logic only if either Charles or Heather goes to the library. Charlie and George do not both go to the library. Either George or Heather or both go to the library. For Heather to go to the library it is necessary that Charlie does not go.*

Use the following atomic propositions:

B: Bob will study logic,
C: Charles goes to the library,
H: Heather goes to the library,
G: George goes to the library.

2.8   Formal Proofs

In this section, we will study a formal proof system called the *natural deduction* system. However, let's first understand the idea behind studying formal proofs. Consider a valid argument such as

Now it is easy to see, either using truth trees or truth tables that this is a valid argument, but this does not tell us anything about the reasoning involved in getting from the hypotheses to the conclusion. Informally this reasoning is as follows: I have $A \longrightarrow B$ and A and so I can get a B here and then I put that together with C to get $B \wedge C$. It is this process that we wish to formally express. There are many formal proof systems, but in this book we will use the Fitch-style natural deduction system (ND). All these formal systems try to formalize the process of reasoning and are especially motivated by the reasoning that mathematicians employ in proving theorems.

Definition 2.8.1 A *Natural Deduction* (ND) proof of a valid argument

$$
\begin{array}{c}
H_1 \\
H_2 \\
\vdots \\
H_n \\
\hline
C
\end{array}
$$

is a finite sequence $A_1, A_2, \ldots, A_m$ of formulas ending with C, (i.e., $A_m = C$) where each formula is either a hypothesis or is obtained from the preceding formulas using an inference rule. We shall denote this proof as

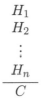

Here the H_i are called the *hypotheses* and C is the *conclusion*. The vertical line to the left of the proof is called the *spine* of the proof and the horizontal line is used to separate the hypotheses and the sequence of formulas constituting the proof.

We say that the hypotheses $H_1, \ldots, H_n$ prove C, denoted $H_1, H_2, \ldots, H_n \vdash C$ if there is a natural deduction proof with the H_i as its hypotheses and C as its conclusion. Expressions of the form $H_1, H_2, \ldots, H_n \vdash C$ are called *sequents*.

The inference rules play the key role in the construction of a proof. In this particular system, inference rules are given as a pair for each logical connective. These are called the *introduction* and *elimination* rules. Intuitively, the introduction rule for a connective, say $\mathcal{C}$ tells you how to construct or prove a formula whose main connective is $\mathcal{C}$, on the other hand, the elimination rule for $\mathcal{C}$, tells you how to deconstruct or to use a formula whose main connective is $\mathcal{C}$. On top of these inference rules to be introduced below, there is one single rule that takes care of formula occurrences at various levels, that is the scope of each formula. In general, a proof will contain various subproofs nested in different groups. This rule tells us that we can use a formula that occurs in a proof inside any of its subproofs.

> **Reiteration:** *Any formula that occurs in a proof can be reiterated in any of its subproofs.*

We next introduce the inference rules for the connectives $\wedge$ and $\longrightarrow$.

$\wedge$-**Introduction** rule:

$$
\begin{array}{c|l}
k & A \\
 & \vdots \\
l & B \\
 & \vdots \\
 & A \wedge B \quad \wedge - I, k, l
\end{array}
\qquad\qquad
\begin{array}{c|l}
k & B \\
 & \vdots \\
l & A \\
 & \vdots \\
 & A \wedge B \quad \wedge - I, k, l
\end{array}
$$

$\wedge$-**Elimination** rule:

$$
\begin{array}{c|l}
k & A \wedge B \\
 & \vdots \\
 & A \quad \wedge - E, k
\end{array}
\qquad\qquad
\begin{array}{c|l}
k & A \wedge B \\
 & \vdots \\
 & B \quad \wedge - E, k
\end{array}
$$

$\longrightarrow$-**Introduction** rule:

$$
\begin{array}{c|l}
\vdots & \\
k & \quad A \\
\vdots & \quad \vdots \\
l & \quad B \\
l+1 & A \longrightarrow B \qquad \longrightarrow - I, k\text{--}l
\end{array}
$$

$\longrightarrow$-**Elimination** rule:

$$
\begin{array}{c|c}
k & A \longrightarrow B \\
 & \vdots \\
l & A \\
 & \vdots \\
 & B \qquad \longrightarrow -E, k, l
\end{array}
\qquad\qquad
\begin{array}{c|c}
k & A \\
 & \vdots \\
l & A \longrightarrow B \\
 & \vdots \\
 & B \qquad \longrightarrow -E, k, l
\end{array}
$$

Here are some examples.

Example 2.8.2 Give a natural deduction proof for $A \wedge B$, $A \wedge C \vdash B \wedge C$.

$$
\begin{array}{c|ll}
1 & A \wedge B & \\
2 & A \wedge C & \\ \hline
3 & B & \wedge - E, 1 \\
4 & C & \wedge - E, 2 \\
5 & B \wedge C & \wedge - I, 3, 4
\end{array}
$$

Example 2.8.3 Give a natural deduction proof for $A, A \longrightarrow B$, $C \vdash B \wedge C$.

$$
\begin{array}{c|ll}
1 & A & \\
2 & A \longrightarrow B & \\
3 & C & \\ \hline
4 & B & \longrightarrow -E, 1, 2 \\
5 & B \wedge C & \wedge - I, 3, 4
\end{array}
$$

Example 2.8.4 Give a natural deduction proof for $A \wedge B \vdash B \wedge A$.

1	$A \wedge B$	
2	A	$\wedge - E,\ 1$
3	B	$\wedge - E,\ 1$
4	$B \wedge A$	$\wedge - I,\ 2,\ 3$

Example 2.8.5 Give a natural deduction proof for $(P \wedge Q) \wedge R \vdash P \wedge (Q \wedge R)$.

1	$(P \wedge Q) \wedge R$	
2	$P \wedge Q$	$\wedge - E,\ 1$
3	R	$\wedge - E,\ 1$
4	P	$\wedge - E,\ 2$
5	Q	$\wedge - E,\ 2$
6	$Q \wedge R$	$\wedge - I,\ 3,\ 5$
7	$P \wedge (Q \wedge R)$	$\wedge - I,\ 4,\ 6$

Example 2.8.6 Give a natural deduction proof for

$$Q, P \wedge T, (P \wedge Q) \longrightarrow S \vdash S.$$

1	Q	
2	$P \wedge T$	
3	$(P \wedge Q) \longrightarrow S$	
4	P	$\wedge - E,\ 2$
5	$P \wedge Q$	$\wedge - I,\ 1,\ 4$
6	S	$\longrightarrow - E,\ 3,\ 5$

Example 2.8.7 Give a natural deduction proof for ⊢ $A \longrightarrow$ $(B \longrightarrow A)$.

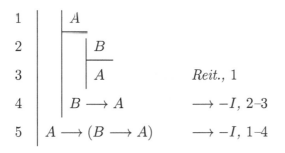

1	A	
2	B	
3	A	*Reit.*, 1
4	$B \longrightarrow A$	$\longrightarrow$ -I, 2–3
5	$A \longrightarrow (B \longrightarrow A)$	$\longrightarrow$ -I, 1–4

Example 2.8.8 Give a natural deduction proof for

$$P \longrightarrow Q, Q \longrightarrow R \vdash P \longrightarrow R.$$

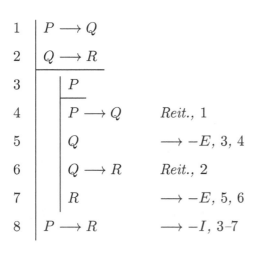

1	$P \longrightarrow Q$	
2	$Q \longrightarrow R$	
3	P	
4	$P \longrightarrow Q$	*Reit.*, 1
5	Q	$\longrightarrow$ -E, 3, 4
6	$Q \longrightarrow R$	*Reit.*, 2
7	R	$\longrightarrow$ -E, 5, 6
8	$P \longrightarrow R$	$\longrightarrow$ -I, 3–7

Example 2.8.9 Give a natural deduction proof for

$$P, Q, (P \wedge Q) \longrightarrow (R \wedge S) \vdash R.$$

1	P	
2	Q	
3	$(P \land Q) \longrightarrow (R \land S)$	
4	$P \land Q$	$\land - I$, 1, 2
5	$R \land S$	$\longrightarrow -E$, 3, 4
6	R	$\land - E$, 5

Example 2.8.10 Give a natural deduction proof for

$$A \longrightarrow (B \longrightarrow C) \vdash (A \land B) \longrightarrow C.$$

1	$A \longrightarrow (B \longrightarrow C)$	
2	$A \land B$	
3	A	$\land - E$, 2
4	$A \longrightarrow (B \longrightarrow C)$	*Reit.*, 1
5	$B \longrightarrow C$	$\longrightarrow -E$, 3, 4
6	B	$\land - E$, 2
7	C	$\longrightarrow -E$, 5, 6
8	$(A \land B) \longrightarrow C$	$\longrightarrow -I$, 2–7

Exercises

Exercise 2.8.1 Give a natural deduction proof for each of the following sequents.

(i) $B \land A \vdash A \land B$,
(ii) $P \land Q, R \land S \vdash Q \land S$,

(iii) $A, A \longrightarrow B, C \vdash B \wedge C,$
(iv) $A \longrightarrow B, B \longrightarrow C \vdash A \longrightarrow C,$
 (v) $(P \longrightarrow Q) \wedge (P \longrightarrow R) \vdash P \longrightarrow (Q \wedge R).$

We shall next introduce the inference rules for negation and disjunction, and then continue with some examples of natural deduction proofs.

¬-Introduction rule:

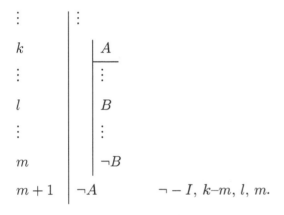

$$m+1 \quad | \quad \neg A \qquad \neg - I, \, k\text{--}m, \, l, \, m.$$

¬-Elimination rule:

$$
\begin{array}{c|l}
k & \vdots \\
& \neg\neg A \\
& \vdots \\
& A \quad \neg - E, k.
\end{array}
$$

∨-Introduction rule:

$$
\begin{array}{c|l}
k & \vdots \\
& A \\
& \vdots \\
& A \vee B \quad \vee - I, k
\end{array}
\qquad
\begin{array}{c|l}
k & \vdots \\
& B \\
& \vdots \\
& A \vee B \quad \vee - I, k.
\end{array}
$$

∨-**Elimination** rule:

$$
\begin{array}{l l l}
\vdots & \vdots & \\
k & A \vee B & \\
k+1 & \quad A & \\
\vdots & \quad \vdots & \\
l & \quad C & \\
l+1 & \quad B & \\
\vdots & \quad \vdots & \\
m & \quad C & \\
m+1 & C & \vee - E,\ k,\ k+1\text{--}l,\ l+1\text{--}m.
\end{array}
$$

Before we look at some more examples, we shall point out that for the only remaining connective, that is bi-implication we replace $A \leftrightarrow B$ by its logical equivalent $(A \longrightarrow B) \wedge (B \longrightarrow A)$.

Example 2.8.2 Give a natural deduction proof for

$$A \longrightarrow D, D \longrightarrow C \vdash (A \wedge B) \longrightarrow C.$$

$$
\begin{array}{l l l}
1 & A \longrightarrow D & \\
2 & D \longrightarrow C & \\
3 & \quad A \wedge B & \\
4 & \quad A & \wedge - E,\ 3 \\
5 & \quad A \longrightarrow D & Reit.,\ 1 \\
6 & \quad D & \longrightarrow - E,\ 4,\ 5 \\
7 & \quad D \longrightarrow C & Reit.,\ 2 \\
8 & \quad C & \longrightarrow - E,\ 6,\ 7 \\
9 & (A \wedge B) \longrightarrow C & \longrightarrow - I,\ 3\text{--}8.
\end{array}
$$

Example 2.8.3 Give a natural deduction proof for
$$P, P \longrightarrow Q, Q \longrightarrow R \vdash R.$$

1	P	
2	$P \longrightarrow Q$	
3	$Q \longrightarrow R$	
4	Q	$\longrightarrow -E,\ 1,\ 2$
5	R	$\longrightarrow -E,\ 3,\ 4\,.$

Example 2.8.4 Give a natural deduction proof for
$$P, Q, (P \wedge Q) \longrightarrow R \vdash R \vee S.$$

1	P	
2	Q	
3	$(P \wedge Q) \longrightarrow R$	
4	$P \wedge Q$	$\wedge - I,\ 1,\ 2$
5	R	$\longrightarrow -E,\ 3,\ 4$
6	$R \vee S$	$\vee - I,\ 5\,.$

Example 2.8.5 Give a natural deduction proof for
$$P \longrightarrow Q, P \longrightarrow \neg Q \vdash \neg P.$$

1	$P \longrightarrow Q$	
2	$P \longrightarrow \neg Q$	
3	P	
4	$P \longrightarrow Q$	*Reit.*, 1
5	$P \longrightarrow \neg Q$	*Reit.*, 2
6	Q	$\longrightarrow -E,\ 3,\ 4$
7	$\neg Q$	$\longrightarrow -E,\ 3,\ 5$
8	$\neg P$	$\neg - I,\ 3\text{--}7,\ 6,\ 7.$

Example 2.8.6 Give a natural deduction proof for

$$Q, P \wedge T, (P \wedge Q) \longrightarrow \neg\neg S \vdash S.$$

1	Q	
2	$P \wedge T$	
3	$(P \wedge Q) \longrightarrow \neg\neg S$	
4	P	$\wedge - E, 2$
5	$P \wedge Q$	$\wedge - I, 1, 4$
6	$\neg\neg S$	$\longrightarrow -E, 3, 5$
7	S	$\neg - E, 6.$

Example 2.8.7 Give a natural deduction proof for $\neg P \longrightarrow Q$, $\neg Q \vdash P$.

1	$\neg P \longrightarrow Q$	
2	$\neg Q$	
3	$\quad\neg P$	
4	$\quad\neg P \longrightarrow Q$	*Reit.*, 1
5	$\quad Q$	$\longrightarrow -E, 3, 4$
6	$\quad\neg Q$	*Reit.*, 2
7	$\neg\neg P$	$\neg - I, 3\text{-}6, 5, 6$
8	P	$\neg - E, 7.$

Example 2.8.8 Give a natural deduction proof for $P \longrightarrow Q, \neg Q \vdash \neg P$.

1	$P \longrightarrow Q$	
2	$\neg Q$	
3	P	
4	$P \longrightarrow Q$	*Reit.*, 1
5	Q	$\longrightarrow -E$, 3, 4
6	$\neg Q$	*Reit.*, 2
7	$\neg P$	$\neg - I$, 3–6, 5, 6.

Example 2.8.9 Give a natural deduction proof for

$$P \longrightarrow Q \vdash \neg Q \longrightarrow \neg P.$$

1	$P \longrightarrow Q$	
2	$\neg Q$	
3	P	
4	$P \longrightarrow Q$	*Reit.*, 1
5	Q	$\longrightarrow -E$, 3, 4
6	$\neg Q$	*Reit.*, 2
7	$\neg P$	$\neg - I$, 3–6, 5, 6
8	$\neg Q \longrightarrow \neg P$	$\longrightarrow -I$, 2–7.

Here are some more examples of natural deduction proofs.

Example 2.8.10 Give a natural deduction proof for $(A \vee B) \longrightarrow C, A \vdash C$.

1	$(A \vee B) \longrightarrow C$	
2	A	
3	$A \vee B$	$\vee - I, 2$
4	C	$\longrightarrow -E, 1, 3.$

Example 2.8.11 Give a natural deduction proof for

$$P \vee R, P \longrightarrow Q, R \longrightarrow Q \vdash Q.$$

1	$P \vee R$	
2	$P \longrightarrow Q$	
3	$R \longrightarrow Q$	
4	$\quad P$	
5	$\quad P \longrightarrow Q$	*Reit.*, 2
6	$\quad Q$	$\longrightarrow -E, 4, 5$
7	$\quad R$	
8	$\quad R \longrightarrow Q$	*Reit.*, 3
9	$\quad Q$	$\longrightarrow -E, 7, 8$
10	Q	$\vee - E, 1, 4\text{--}6, 7\text{--}9.$

Example 2.8.12 Give a natural deduction proof for $\neg(P \vee Q) \vdash \neg P \wedge \neg Q$.

1	$\neg(P \vee Q)$	
2	$\quad P$	
3	$\quad P \vee Q$	$\vee - I, 2$
4	$\quad \neg(P \vee Q)$	$Reit., 1$
5	$\neg P$	$\neg - I, 2\text{-}4,\ 3,\ 4$
6	$\quad Q$	
7	$\quad P \vee Q$	$\vee - I, 6$
8	$\quad \neg(P \vee Q)$	$Reit., 1$
9	$\neg Q$	$\neg - I, 6\text{-}8,\ 7,\ 8$
10	$\neg P \wedge \neg Q$	$\wedge - I, 5,\ 9.$

Example 2.8.13 Give a natural deduction proof for

$$(A \vee B) \longrightarrow C \vdash A \longrightarrow C.$$

1	$(A \vee B) \longrightarrow C$	
2	$\quad A$	
3	$\quad A \vee B$	$\vee - I, 2$
4	$\quad (A \vee B) \longrightarrow C$	$Reit., 1$
5	$\quad C$	$\longrightarrow - E, 3,\ 4$
6	$A \longrightarrow C$	$\longrightarrow - I, 2\text{-}5.$

Example 2.8.14 Give a natural deduction proof for $(B \wedge A) \vee (A \wedge C) \vdash A$.

1	$(B \wedge A) \vee (A \wedge C)$	
2	$\quad B \wedge A$	
3	$\quad A$	$\wedge - E, 2$
4	$\quad A \wedge C$	
5	$\quad A$	$\wedge - E, 4$
6	A	$\vee - E, 1, \text{2-3}, \text{4-5}.$

Example 2.8.15 Give a natural deduction proof for $\neg P \longrightarrow Q \vdash P \vee Q$.

1	$\neg P \longrightarrow Q$	
2	$\quad \neg(P \vee Q)$	
3	$\quad\quad P$	
4	$\quad\quad P \vee Q$	$\vee - I, 3$
5	$\quad\quad \neg(P \vee Q)$	$Reit., 2$
6	$\quad \neg P$	$\neg - I, \text{3-5}, 4, 5$
7	$\quad \neg P \longrightarrow Q$	$Reit., 1$
8	$\quad Q$	$\longrightarrow - E, 6, 7$
9	$\quad P \vee Q$	$\vee - I, 8$
10	$\neg\neg(P \vee Q)$	$\neg - I, \text{2-9}, 2, 9$
11	$P \vee Q$	$\neg - E, 10.$

Example 2.8.16 Give a natural deduction proof for $\vdash A \longrightarrow \neg\neg A$.

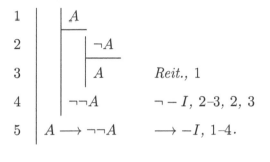

1	A	
2	$\neg A$	
3	A	*Reit.*, 1
4	$\neg\neg A$	$\neg - I$, 2–3, 2, 3
5	$A \longrightarrow \neg\neg A$	$\longrightarrow -I$, 1–4.

Remark 2.8.17

- *In general a proof follows a sequence of alternating bottom-up (introduction rules) and top-down (elimination rules) strategies. A good hint though is to always consider the bottom-up strategy at the beginning.*
- *To prove $A \longrightarrow B$ first try the $\longrightarrow -I$ rule.*
- *To prove $\neg A$ first try $\neg - I$ rule.*
- *To prove $A \vee B$, first try $\neg - I$ rule. Unless it is obvious that either a top-down strategy, or a strategy using $\vee - I$ rule exists.*
- *The inference rules have the very important property that they preserve the truth. That is to say, assuming that every formula above the point of application of a rule is true, so is the formula obtained after the application of the rule. Hence, it is not possible to have a rule that gets false propositions from true ones. Therefore, rules like getting A or B from $A \longrightarrow B$ are WRONG. Same for getting A or B from $A \vee B$. This is because, for example A can be false, when $A \longrightarrow B$ is true, etc.*

At this point we might wish to pause and think about what we have done so far. We started by introducing an artificial language called the propositional logic. It was not very powerful but we still could take arguments in English and translate them into this language and check their validity. Or we could check the consistency of a specification. So, on the one hand, we have this world of arguments, true and false, and tautologies, etc. On the other hand, we studied formal proof systems, namely the Natural Deduction System. We can now devise ND proofs for provable sequents.

A natural question arises: Is there a relationship between these two worlds, that is the world of true and false and the world of formal proofs? Indeed, there is a relationship and it is a very tight one. It says that you can prove *all and only* the valid arguments. That is, you can prove *only* valid arguments (**Soundness**) and you can prove *all* valid arguments (**Completeness**). Formally,

Theorem 2.8.18 (Completeness) *Let* $A_1, \ldots, A_n$ *be a sequence of formulas and* C *be a formula. Then*

$$A_1, \ldots, A_n \vdash C \quad \textit{iff} \quad \begin{array}{c} A_1 \\ \vdots \\ \underline{A_n} \\ C \end{array} \quad \textit{is a valid argument.}$$

So, in particular, it says that $\vdash C$ iff C is a tautology. Now you have learnt another method for checking that a formula is a tautology and also you have learnt another method for checking the validity of an argument. For example, now you can show that $A \longrightarrow (B \longrightarrow A)$ is a tautology by just trying to prove it with no hypothesis, recall that we did this proof as one of our examples above.

We will next try to enrich our formal (artificial) language so that we can have more expressive power. We shall do so by adding variables and connectives that control variables. This will give us tremendous expressive power over propositional language, however we pay the price, in that validity, truth, proofs, etc. will become more complex. In order to understand the necessary structures in our new language we shall briefly review some set theory and recall the definitions of functions and relations.

Exercises

Exercise 2.8.2 Give a natural deduction proof for the following sequents.

1. $(A \vee B) \longrightarrow C \vdash A \longrightarrow C$,
2. $A \longrightarrow (B \vee D), B \longrightarrow C \vdash A \longrightarrow (C \vee D)$,
3. $X \longrightarrow Y, Y \longrightarrow Z, \neg Z \vdash \neg X$,
4. $A \wedge \neg B \vdash \neg (A \wedge B)$,
5. $(P \wedge R) \longrightarrow (\neg\neg Q \wedge S), R \vdash P \longrightarrow Q$,
6. $\vdash P \longrightarrow (Q \longrightarrow P)$,
7. $(B \wedge A) \vee (A \wedge C) \vdash A$,
8. $P \vee R, P \longrightarrow Q, R \longrightarrow Q \vdash Q$,
9. $P \vee Q, \neg Q \vdash P$,
10. $(B \longrightarrow A) \wedge (A \vee B) \vdash A$.

Exercise 2.8.3 Decide if $A \wedge B$ is provable from the hypotheses $A, B \longrightarrow C, \neg C$. If it is provable, then give a natural deduction proof. Otherwise, show that it is not provable.

Chapter 3

Set Theory

3.1 Set Concepts

Set theory was invented by Georg Cantor in 1880s to provide a unified language for mathematics. Set theory is an important branch of mathematics and it has found numerous applications in computing and information sciences. For example, sets are part of the mathematical foundations of data structures such as hash tables, graphs, and trees. They also show up in various modelings of data and databases.

Let us start with some very basic definitions.

Definition 3.1.1 (Set) A *set* is a collection of objects. The objects that make up the set are called its *members*.

As it is clear from the definition of a set, the primary notion in set theory is that of *membership*, thus order of elements or their multiplicity do not matter.

Sets can be represented in different ways, i.e., set notations. The two most common set notations are the set roster notation and set builder notation.

Definition 3.1.2 (Set roster notation) The *set roster notation* is the mathematical form that lists all the members of the set inside the curly braces.

We use left and right curly braces to denote a set. For example $A = \{a, b, c, d\}$ uses the set roster notation to represent a set whose members are the letters $a, b, c,$ and d and nothing else.

Listing all the elements of a set is only possible when the set does not have many elements. If the set contains a lot of elements, we can use ellipsis (...) to omit some elements if the pattern of the elements is obvious and omitting some elements will not lead to confusion or ambiguity.

In the following examples, represent the sets using the roster notation.

Example 3.1.3 The set V of all vowels in the English alphabet.

Solution: $V = \{a, e, i, o, u\}$.

Example 3.1.4 The set $\mathbb{N}$ of all natural numbers.

Solution: $\mathbb{N} = \{0, 1, 2, 3, \ldots\}$.

Example 3.1.5 The set A of all lower case letters in the English alphabet.

Solution: $A = \{a, b, c, \ldots, z\}$.

Definition 3.1.6 (Set builder notation) In this notation, we define a set by specifying the properties of all its members.

The set builder notation builds a set by describing its members. The general notation is $S = \{x \mid P(x)\}$, where S is a set; x is any element of the set; and $P(x)$ is the property of set members. For example, $S = \{x \mid x \in \mathbb{N}, \ x > 0 \text{ and } x < 5\}$ defines the set S that contains $1, 2, 3$, and 4 and only these as members, i.e., $S = \{1, 2, 3, 4\}$.

In the following examples, use the set builder notation to describe each set.

Example 3.1.7 $A = \{2, 4, 6, 8, 10\}$.

Solution: $A = \{x \mid x \text{ is an even natural number and } 0 < x \leq 10\}$.

Example 3.1.8 $B = \{2, 3, 4, 5, 6\}$.

Solution: $B = \{x \mid x \in \mathbb{N} \wedge 2 \leq x \leq 6\}$.

Example 3.1.9 $D = \{0, 1\}$.

Solution: $D = \{x \mid x \in \mathbb{N} \text{ and } x = x^2\}$.

Note that there may be more than one way to specify the members of a set. For example, in the last example above we could have used the property, "Natural number x such that $0 \le x \le 1$" to define the same set. There are many other ways as well.

If an object belongs to a set, the object must satisfy the properties of the set, and conversely if an object satisfies the defining properties of a set, the object must be a member of the set. Given the set $A = \{a, b, c, d\}$, we use the notation $a \in A$ to show that a is a member of the set A, and we use $e \notin A$ to show that e is not a member of the set A.

Example 3.1.10 Given the set $S = \{\{1, 2\}, 3, 4, 5\}$, determine if each statement below is true or false.

1. $3 \in S$.
2. $2 \in S$.
3. $\{1, 2\} \in S$.

Solution: (1) *is true,* (2) *is false, and* (3) *is true.*

We will use the following standard notations to refer to the sets below.

- **The set of natural numbers:** $\mathbb{N} = \{0, 1, 2, 3, \ldots\}$.
- **The set of integers:** $\mathbb{Z} = \{\ldots, -3, -2, -1, 0, 1, 2, 3, \ldots\}$.
- **The set of rational numbers:** $\mathbb{Q} = \left\{ \frac{a}{b} \mid a, b \in \mathbb{Z} \text{ and } b \ne 0 \right\}$.
- **The set of real numbers:** $\mathbb{R}$ which is the set of all rational and irrational numbers.
- **Empty set or null set:** $\emptyset$ or $\{\}$. This set has no elements.
- **Universal set:** U. It is also called the *universe*. This set contains all the sets that we talk about in a given specific context.

Next we introduce a very useful and convenient pictorial language for set operations, called *Venn Diagrams.*

A *Venn diagram*, also known as set diagram or logical diagram, was invented by British logician John Venn in 1881. It uses overlapping shapes (circles, ovals, rectangles, etc.) to illustrate the logical relationships between two or more sets. A Venn diagram is a geometric (pictorial) way of visualizing sets. Venn diagrams are very useful in picturing sets and set operations. However, they are not a substitute for a proof involving sets. The picture below shows two sets A and B with some overlap.

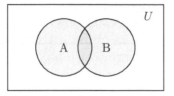

Now that we have looked at the definition of sets, let us proceed by looking at the ways sets interact with each other.

Definition 3.1.11 (Subset) A set A is a *subset* of a set B, denoted $A \subseteq B$, if every element of A is also an element of B.

For example, the set $A = \{a, b, c\}$ is a subset of the set $B = \{a, b, \ldots, z\}$. However, B is not a subset of A, denoted $B \nsubseteq A$, because, for example the element $e \in B$ does not belong in A, i.e., $e \notin A$.

Definition 3.1.12 (Proper subset) A set A is a *proper subset* of a set B, denoted $A \subset B$, if A is a subset of B and A is not equal to B. Sometimes we use the notation $A \subsetneq B$.

In the above example, the set $A = \{a, b, c\}$ is also a proper subset of the set $B = \{a, b, \ldots, z\}$ because $A \subseteq B$ and A is not equal to B. According to the definitions of subset and proper subset, we can see that any set is a subset of itself; however it is not a proper subset of itself. That is $A \subseteq A$, but $A \not\subset A$.

Venn diagrams for subsets and proper subsets:

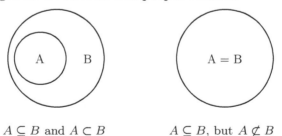

$A \subseteq B$ and $A \subset B$ $A \subseteq B$, but $A \not\subset B$

As we discussed in the previous chapter, once a mathematical object is introduced, mathematicians are interested in defining an appropriate notion of sameness between these objects. For example, in the previous chapter we looked at the sameness of propositional formulas that was captured with the notion of logical equivalence. What about sets?

Definition 3.1.13 (Set equality) Two sets A and B are said to be equal, denoted $A = B$, if and only if $A \subseteq B$ and $B \subseteq A$, meaning that all elements of A belong in B and all elements of B belong in A. When the two sets do not have the same elements, they are said to be not equal, denoted $A \neq B$.

Example 3.1.14 Given three sets A, B and C: $A = \{2, 4, 6, 8\}, B = \{1, 2, \ldots, 10\}$, and $C = \{x \mid x$ is an even natural number and $0 < x < 10\}$. Determine if each of the following is true or false.

1. $A \subseteq B$.
2. $A \subset B$.
3. $A \subseteq C$.
4. $A \subset C$.
5. $A = C$.
6. $B \neq C$.

Solution: (1) *is true,* (2) *is true,* (3) *is true,* (4) *is false,* (5) *is true, and* (6) *is true.*

A set A is said to be *finite* if it has finitely many members, otherwise it is said to be an *infinite* set. For example, $\{1, 2, 3, 4\}$ is a finite set, whereas $\mathbb{N}$ is an infinite set.

Definition 3.1.15 (Cardinality) For a finite set A, we define its *cardinality*, denoted $|A|$ to be the number of elements in A.

For example, let $A = \{x \mid x \text{ odd}, x \in \mathbb{N}, x < 10\}$, then $|A| = 5$. Let $B = \emptyset$, then $|B| = 0$.

We shall not discuss the notion of cardinality for infinite sets in this book.

There are various situations when it is useful to consider the set containing all the subsets of a given set.

Definition 3.1.16 (Power set) Given a set A, the *power set* of A, denoted by $\mathcal{P}(A)$, is the set of all subsets of A.

For example, let $A = \{0, 1\}$, then $\mathcal{P}(A) = \{\emptyset, \{0\}, \{1\}, \{0, 1\}\}$. Let $B = \emptyset$, then $\mathcal{P}(B) = \{\emptyset\}$. If $C = \{\emptyset\}$, then $\mathcal{P}(C) = \{\emptyset, \{\emptyset\}\}$. It is a fact that if $|A| = n$, then $|\mathcal{P}(A)| = 2^n$. You can easily verify this formula in the special cases above. Please note that $\emptyset$ is an element of any power set.

Example 3.1.17 Let $A = \{x, y, \{x, y\}\}$. Find the following:

1. $|A|$.
2. $|\mathcal{P}(A)|$.
3. $\mathcal{P}(A)$.

Solution:

1. $|A| = 3$.
2. $|\mathcal{P}(A)| = 2^3 = 8$.
3. $\mathcal{P}(A) = \{$

$$\emptyset, \tag{1}$$

$$\{x\}, \{y\}, \{\{x, y\}\}, \tag{2}$$

$$\{x, y\}, \{x, \{x, y\}\}, \{y, \{x, y\}\}, \tag{3}$$

$$\{x, y, \{x, y\}\}, \tag{4}$$

$\}$.

Definition 3.1.18 An *ordered n-tuple* $(a_1, a_2, \ldots, a_n)$ is an ordered list with a_1 the first, a_2 the second, $\ldots$, and a_n the nth element.

A tuple is also known as a list. A tuple is often written by listing all the elements within parentheses, separated by commas. The two important properties that distinguish a tuple from a set are:

1. A tuple allows and recognizes repetition of elements so $(a, b, b, c) \neq (a, b, c)$, whereas $\{a, b, b, c\} = \{a, b, c\}$.
2. Tuple elements are ordered so $(a, b, c) \neq (a, c, b)$, whereas $\{a, b, c\} = \{a, c, b\}$.

Because the order of elements in a tuple matters, two ordered tuples $(a_1, \ldots, a_n)$ and $(b_1, \ldots, b_n)$ are *equal* iff $a_i = b_i$ for all $i = 1, 2, \ldots, n$. A tuple with two elements is called an ordered *pair*. For examples, (a, b) and $(1, 1)$ are ordered pairs.

Definition 3.1.19 (Cartesian product of sets) The Cartesian product of the sets $A_1, A_2, \ldots A_n$, denoted by $A_1 \times A_2 \times \cdots \times A_n$, is defined as

$$A_1 \times A_2 \times \cdots \times A_n = \{(a_1, a_2, \ldots, a_n) \mid a_i \in A_i \text{ for } i = 1, \ldots, n\}.$$

For example, let $A_1 = \{a, b\}$ and $A_2 = \{1, 2\}$. Then

$$A_1 \times A_2 = \{(a, 1), (a, 2), (b, 1), (b, 2)\}.$$

Given finite sets A_i, $i = 1, \ldots, n$, it can be easily seen that

$$|A_1 \times \cdots \times A_n| = |A_1| \times \cdots \times |A_n|.$$

Example 3.1.20 Let A_1, A_2, and A_3 be the sets, $A_1 = \{a, b\}$, $A_2 = \{1, 2\}$, and $A_3 = \{0, 1, 2\}$. Find the following:

1. $|A_1 \times A_2 \times A_3|$.
2. $A_1 \times A_2 \times A_3$.

Solution:

1. $|A_1 \times A_2 \times A_3| = 2 \times 2 \times 3 = 12$.
2. $\quad A_1 \times A_2 \times A_3 = \{$

$$(a, 1, 0), \tag{1}$$

$$(a, 1, 1), \tag{2}$$

$$(a, 1, 2), \tag{3}$$

$$(a, 2, 0), \tag{4}$$

$$(a, 2, 1), \tag{5}$$

$$(a, 2, 2), \tag{6}$$

$$(b, 1, 0), \tag{7}$$

$$\ldots \ldots$$

$$(b, 2, 2) \tag{12}$$

$$\}.$$

Example 3.1.21 Let $A = \{0, 1\}$ and $B = \{\triangle\}$. Find the following:

1. $|A \times B \times A|$.
2. $A \times B \times A$.

Solution:

1. $|A \times B \times A| = 2 \times 1 \times 2 = 4$.
2. $\quad A \times B \times A = \{$

$$(0, \triangle, 0), \tag{1}$$
$$(0, \triangle, 1), \tag{2}$$
$$(1, \triangle, 0), \tag{3}$$
$$(1, \triangle, 1), \tag{4}$$
$$\}.$$

Exercises

Exercise 3.1.1 Given the sets $A = B = \{0, 1\}$, find $A \times B$.

Exercise 3.1.2 Given the sets $A = B = \{0, 1\}$, find $\mathcal{P}(A) \times B$.

Exercise 3.1.3 Given the sets $A = \{0, 1\}$, and $B = \{3\}$, is $\mathcal{P}(B) \subseteq A$?

Exercise 3.1.4 Let $A = \{x \mid x$ is an even prime number$\}$. Is $3 \in A$?

Exercise 3.1.5 Let $A = \{x \in \mathbb{N} \mid x \leq 10\}$. Find $|\mathcal{P}(A) \times A|$.

Exercise 3.1.6 Let $D = \{0, 1, 2, 3, 6, 5, 10, 11, 12, 15\}$

1. Find three subsets of D.
2. Use the set-builder notation to describe $\{2, 3, 5, 11\}$.
3. Let $S = \{3, 6, 12, 15\}$ and $T = \{x \mid x \in D$ and x is divisible by 3$\}$. Answer the following questions:
   a. Is $S \subseteq T$?
   b. Is $S = T$?
   c. Is $S \subset T$?
   d. Find a set of which S is a proper subset.

Exercise 3.1.7 Given the set $E = \{Peter, Kim, Mike\}$, determine the following:

1. $|\mathcal{P}(E)|$.
2. $\mathcal{P}(E)$.

Exercise 3.1.8 Given the set $F = \{\emptyset, 0, \{0, 1\}\}$, determine the following:

1. $|\mathcal{P}(F)|$.
2. $\mathcal{P}(F)$.

Exercise 3.1.9 Given the sets $S = \{Peter, Kim\}, T = \{Frank, Henry\}$, and $M = \{Mary, Kevin\}$. Determine the following:

1. $|S \times T|$.
2. $|S \times T \times M|$.
3. $S \times T$.
4. $S \times T \times M$.

3.2   Set Operations

When dealing with multiple sets, we often need to determine how their elements are related. For example, we may need to know, given two sets, what elements they have in common, or what elements are unique to one specific set. There comes the concept of set operations. Set operations refer to the operations that are applied to two or more sets to build new sets. There are four main set operations, namely set union, set intersection, set difference, and set complement.

Definition 3.2.1 (Set union) The *union* of the sets A and B, denoted by $A \cup B$, is the set containing all the elements that are either in A, or B, or both. That is,

$$A \cup B = \{x \mid x \in A \vee x \in B\}.$$

In the following Venn diagram, the colored area in teal represents the union of the sets A and B.

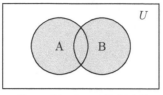

Venn diagram of $A \cup B$

Definition 3.2.2 (Set intersection) The *intersection* of the sets A and B, denoted by $A \cap B$, is the set containing all the elements that are in both A and B. That is,

$$A \cap B = \{x \mid x \in A \wedge x \in B\}.$$

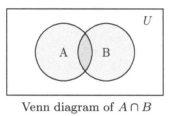

Venn diagram of $A \cap B$

Definition 3.2.3 (Set difference) Set A minus set B, denoted by $A - B$, is the set containing all the elements that are in A, but not in B. That is,

$$A - B = \{x \mid x \in A \wedge x \notin B\}.$$

Please note $A - B \neq B - A$.

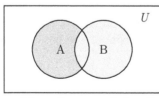 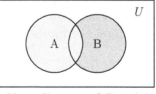

Venn diagram of $A - B$ Venn diagram of $B - A$

Definition 3.2.4 (Set complement) The *complement* of a set A (with respect to the universe U), denoted by $\overline{A}$ or A^c, is the set containing all the elements that are not in A. That is,

$$\overline{A} = U - A = \{x \mid x \notin A\}.$$

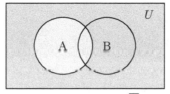

Venn diagram of $\overline{A}$

Example 3.2.5 Given three sets $A = \{1, 2, 3\}$, $B = \{3, 4, 5\}$, and $U = \{1, 2, \ldots, 10\}$, find each of the following:

1. $A \cup B$.
2. $A \cap B$.
3. $A - B$.
4. $B - A$.
5. $\overline{A}$.

Solution:

1. $A \cup B = \{1, 2, 3, 4, 5\}$.
2. $A \cap B = \{3\}$.
3. $A - B = \{1, 2\}$.
4. $B - A = \{4, 5\}$.
5. $\overline{A} = \{4, 5, 6, 7, 8, 9, 10\}$.

Example 3.2.6 Let $A = \{1, 2, 3, 4\}, B = \{1, 2\}, U = \mathbb{N}$. Find the following:

1. $A \cup B$.
2. $A \cap B$.
3. $A \cap U$.
4. $A - B$.
5. $B - A$.

Solution:

1. $A \cup B = \{1, 2, 3, 4\}$.
2. $A \cap B = \{1, 2\}$.
3. $A \cap U = A = \{1, 2, 3, 4\}$.
4. $A - B = \{3, 4\}$.
5. $B - A = \emptyset$.

Exercises

Exercise 3.2.1 Show that in general, for sets A and B, $A - B \neq B - A$. Can you find specific sets A and B for which $A - B = B - A$?

Exercise 3.2.2 Given sets $U, A,$ and B: $U = \{0, 1, 2, 3, 4, 5, 6, 7, 8, 9\}$, $A = \{x | x \in U, x$ is divisible by $2\}$, and $B = \{x | x \in U, x$ is divisible by $4\}$. Determine the following:

1. $A \cup B$.
2. $A \cap B$.
3. $B - A$.
4. $A - B$.
5. A^c.

Exercise 3.2.3 Given any set A and the universal set U, determine the results of the following operations:

1. $A \cup \emptyset$.
2. $A \cap \emptyset$.
3. $A \cap U$.
4. $A \cup U$.
5. $A \cup A^c$.
6. $A \cap A^c$.

3.3   Set Theoretical Relationships

In this section, we will discuss how to prove and disprove two types of relationships: set identities and subset relations.

Let us first observe some logical relationships related to the set operations we just defined.

1. $x \in A \cup B \iff (x \in A) \vee (x \in B)$.
2. $x \in A \cap B \iff (x \in A) \wedge (x \in B)$.
3. $x \in A - B \iff (x \in A) \wedge (x \notin B)$.
4. $x \in A^c \iff x \notin A$.
5. $x \notin A \cup B \iff (x \in A^c) \wedge (x \in B^c)$.
6. $x \notin A \cap B \iff (x \in A^c) \vee (x \in B^c)$.

The last two equivalences follow from the definition of intersection and union and De Morgan laws.

So, given two sets A and B, how shall we prove that A is a subset of B? Recall that, in order to do this we have to show that any element of A, let us call it x, must also be a member of B. This must hold no matter what element of A we pick. So we pick an element x in the universe U, and we assume that x is in A, and then we try to show that x is also an element of B. That is, we try to show that the formula below is true.

$$\text{For any } x \in U, (x \in A \longrightarrow x \in B).$$

Example 3.3.1 Show that $A \cap B \subseteq A$.

Solution:

We pick an arbitrary element $x \in U$ and assume that $x \in A \cap B$. We need to show $x \in A$.

1. $x \in A \cap B$ *assumption,*
2. $x \in A \wedge x \in B$ *by definition of $\cap$,*
3. $x \in A$ *logical consequence of step (2) above.*

The three steps above show that any element of $A \cap B$ is also an element of A.

Example 3.3.2 Show that $A \subseteq A \cup B$.

Solution:

We pick an arbitrary element $x \in U$, and assume that $x \in A$, we need to show $x \in A \cup B$.

1. $x \in A$ *assumption,*
2. $x \in A \vee x \in B$ *logical consequence of step (1),*
3. $x \in A \cup B$ *by definition of $\cup$.*

The three steps above show that any element of A is also an element of $A \cup B$.

Example 3.3.3 Show that $A \cap B \subseteq A \cup B$.

Solution:

We pick an arbitrary element $x \in U$, assume that $x \in A \cap B$ and try to show that it also belongs to $A \cup B$.

    1. $x \in A \cap B$ *assumption,*
    2. $x \in A \wedge x \in B$ *by definition of $\cap$,*
    3. $x \in A$ *logical consequence of (2),*
    4. $x \in A \vee x \in B$ *logical consequence of (3),*
    5. $x \in A \cup B$ *by definition of $\cup$.*

    The five steps above show that any element of $A \cap B$ is also an element of $A \cup B$.

Example 3.3.4 Show that $A \cap C \subseteq A - (B - C)$.

Solution:

Let us first see what the set on the right hand side looks like.

$$
\begin{aligned}
A - (B - C) &= \{x \,|\, x \in A \wedge \neg(x \in (B - C))\} \\
&= \{x \,|\, x \in A \wedge \neg(x \in B \wedge x \notin C)\} \\
&= \{x \,|\, x \in A \wedge (x \notin B \vee x \in C)\} \\
&= \{x \,|\, (x \in A \wedge x \notin B) \vee (x \in A \wedge x \in C)\} \\
&= \{x \,|\, (x \in A \wedge x \in \overline{B}) \vee (x \in A \wedge x \in C)\} \\
&= (A \cap \overline{B}) \cup (A \cap C).
\end{aligned}
$$

    1. $x \in A \cap C$ *assumption,*
    2. $x \in (A \cap C) \vee x \in (A \cap \overline{B})$ *logical consequence of (1),*
    3. $x \in (A \cap C) \cup (A \cap \overline{B})$ *by definition of $\cup$.*
    4. $x \in A - (B - C)$ *by the identity above.*

    These steps show any element of $A \cap C$ is also an element of $A - (B - C)$.

    A set identity is an equation that is universally true. Here are some examples:

- **Commutativity of ∩:** $A \cap B = B \cap A$.
- **Associativity of ∩:** $A \cap (B \cap C) = (A \cap B) \cap C$.
- **Distributivity of ∩ over ∪:** $A \cap (B \cup C) = (A \cap B) \cup (A \cap C)$.

One way to show a set identity is to use the definition of set equality. That is, we show that $A = B$ by showing that $A \subseteq B$ and $B \subseteq A$.

Here is an example.

Example 3.3.5 Show that $A - B = A \cap B^c$.

Solution:

The solution consists of two subproofs. In the first subproof, we show $A - B \subseteq A \cap B^c$. In the second subproof, we show $A \cap B^c \subseteq A - B$. Here is the first subproof:

1. $x \in A - B$	*assumption,*
2. $x \in A \land x \notin B$	*by definition of set difference,*
3. $x \in A \land x \in B^c$	*by definition of set compliment,*
4. $x \in A \cap B^c$	*by definition of set intersection.*

These four steps show that any element of $A - B$ is also an element of $A \cap B^c$.

Here is the second subproof:

1. $x \in A \cap B^c$	*assumption,*
2. $x \in A \land x \in B^c$	*by definition of set intersection,*
3. $x \in A \land x \notin B$	*by definition of set compliment,*
4. $x \in A - B$	*by definition of set difference.*

These four steps show that any element of $A \cap B^c$ is also an element of $A - B$. And the two subproofs together show $A - B = A \cap B^c$.

In the next examples, we use set builder notation and logical equivalences to show two sets are equal.

Example 3.3.6 Show that $A - B = A \cap \overline{B}$.

Solution:

$$
\begin{aligned}
A - B &= \{x \mid x \in A \wedge x \notin B\} \\
&= \{x \mid x \in A \wedge x \in \overline{B}\} \\
&= \{x \mid x \in A \cap \overline{B}\} \\
&= A \cap \overline{B}.
\end{aligned}
$$

Example 3.3.7 Show that $A \cap U = A$.

Solution:

Recall that all sets are subsets of the universal set, that is $A \subseteq U$, therefore any $x \in A$ is also in U.

$$
\begin{aligned}
A \cap U &= \{x \mid x \in A \wedge x \in U\} \\
&= \{x \mid x \in A\} \\
&= A.
\end{aligned}
$$

Example 3.3.8 Show that $\overline{A \cup B} = \overline{A} \cap \overline{B}$.

Solution:

$$
\begin{aligned}
\overline{A \cup B} &= \{x \mid \neg(x \in (A \cup B))\} \\
&= \{x \mid \neg(x \in A \vee x \in B)\} \\
&= \{x \mid (x \notin A \wedge x \notin B)\} \\
&= \{x \mid (x \in \overline{A} \wedge x \in \overline{B})\} \\
&= \{x \mid x \in \overline{A} \cap \overline{B}\} \\
&= \overline{A} \cap \overline{B}.
\end{aligned}
$$

3.3.1 *Disproof of subset relations and set identities*

To disprove a statement means to show that the statement is false by finding a counterexample or contradiction. To disprove a subset relation $A \subseteq B$, we need to find an element $x \in A$ and show $x \notin B$. To disprove a set identity $A = B$, we need to find an element $x \in A$ and show $x \notin B$ or find an element $x \in B$ and show $x \notin A$.

Example 3.3.9 Disprove: For any sets A and B, $A - B \neq \emptyset$.

Solution:

To disprove the statement, we need to find two sets A and B so that $A - B = \emptyset$. We know when two sets are equal, their difference is equal to an empty set. So there are many solutions to the problem. Here is one counterexample:

Let $A = \{1, 2\}$ and $B = \{1, 2\}$.
So $A - B = \emptyset$.

We also know if $A \subseteq B$, then $A - B = \emptyset$. So here is another counterexample:

Let $A = \{a\}$ and $B = \{a, b, c\}$.
So $A - B = \emptyset$.

Example 3.3.10 Disprove: If $A \subseteq B$, then $B - A \neq \emptyset$.

Solution:

This is a conditional statement which translates into an implication in propositional logic. To disprove or make an implication false, we need to make the antecedent of the implication true and its consequent false. We need to find two sets A and B so that $A \subseteq B$, but $B - A = \emptyset$. We know that any set is a subset of itself and the difference between a set and itself is the empty set. So a valid counterexample could consist of two sets that are equal. There are many solutions to the problem. Here is one valid counterexample:

Let $A = \{2, 5\}$ and $B = \{2, 5\}$.
So $A \subseteq B$ is true, but $B - A \neq \emptyset$ is false.

Example 3.3.11 Disprove: If $A \cap B = B \cap C$, then $A = B$.

Solution:

To disprove the statement, we need to find three sets A, B, and C such that $A \cap B = B \cap C$ is true, but $A = B$ is false. Here is one valid counterexample:

Let $A = \{1\}, B = \{2\}$, and $C = \{3\}$.
Then $A \cap B = B \cap C = \emptyset$, but $A \neq B$.

Finally, it is worth mentioning that there is a nice correspondence between propositional logic and set theory that is summarized in the dictionary below.

Sets	Formulas
A	A
$\emptyset$	F
U	T
$\overline{A}$	$\neg A$
$A \cap B$	$A \wedge B$
$A \cup B$	$A \vee B$
$A \subseteq B$	$A \longrightarrow B$ is a tautology
$A = B$	$A \equiv B$

Exercises

Exercise 3.3.1 Given any two sets A and B, show that $A - B \subseteq A$.

Exercise 3.3.2 Disprove the statement: If $A \cup B = B \cap C$, then $A = B$.

Exercise 3.3.3 Given any two sets A and B, is it true that $A \cup B \subseteq A \cap B$?

Exercise 3.3.4 Can you find non-empty sets A, B such that $A \cup B \subseteq A \cap B$?

3.4   Functions and Relations

In this section, we briefly introduce the mathematical notions of function and relation. We shall further study these topics in a later chapter of this book in more detail.

3.4.1   *Relations*

Let's use an example to show what relations are in mathematics. Let $A = \{0, 1, 2\}$ and $B = \{1, 2, 3\}$. Given any elements $x \in A$ and $y \in B$, let us say that x related to y in the relation R iff x is less than y. Consider the following three questions:

1. What is the Cartesian product $A \times B$?
2. What is the set that contains all the pairs related by R?
3. What is the set that contains all the pairs not related by R?

Given the two sets A and B, and the definition of the relation R between A and B, we can construct the following table:

$A \times B$		Related $x < y$		Unrelated $x \geq y$	
x	y	x	y	x	y
0	1	0	1		
0	2	0	2		
0	3	0	3		
1	1			1	1
1	2	1	2		
1	3	1	3		
2	1			2	1
2	2			2	2
2	3	2	3		

The table above shows that the set of $A \times B$ contains 9 ordered pairs. Among them, only 6 are related. The other 3 are unrelated. We can think of the relation R between sets A and B as a subset of $A \times B$ that contains the 6 related pairs.

Definition 3.4.1 (Relation) Let A and B be sets. A *relation* R from A to B is a subset of $A \times B$. Given an ordered pair (x, y) in $A \times B$, we say that x and y are related via R iff (x, y) is in R. Set A is called the *domain* of R, and set B is called the *co-domain* of R.

Example 3.4.2 Let $A = \{1, 2\}$ and $B = \{1, 2, 3\}$. Define a relation R from A to B as follows: Given any $(x, y) \in A \times B$,

$$(x, y) \in R \text{ iff } \frac{x - y}{2} \text{ is an integer.}$$

Answer the following questions:

1. List all members of R.
2. What are the domain and co-domain of R?

Solution:

Let's list all the ordered pairs in $A \times B$ and determine if each pair is in R:

1. $(1,1) \in R$ because $\dfrac{1-1}{2} = 0$ *is an integer.*

   $(1,2) \notin R$ because $\dfrac{1-2}{2} = -\dfrac{1}{2}$ *is not an integer.*

   $(1,3) \in R$ because $\dfrac{1-3}{2} = -1$ *is an integer.*

   $(2,1) \notin R$ because $\dfrac{2-1}{2} = \dfrac{1}{2}$ *is not an integer.*

   $(2,2) \in R$ because $\dfrac{2-2}{2} = 0$ *is an integer.*

   $(2,3) \notin R$ because $\dfrac{2-3}{2} = -\dfrac{1}{2}$ *is not an integer.*

   Therefore, $R = \{(1,1), (1,3), (2,2)\}$.
2. *The domain of R is $A = \{1,2\}$ and the co-domain of R is $B = \{1,2,3\}$.*

The notion of a relation can be extended to any number of sets. For example, the set $R = \{(1,1,1), (2,2,2)\}$ defines a relation on $A \times B \times A$ with A and B as above. Similarly, the set $R = \{(1,2,1,2), (2,2,2,2), (1,1,2,2)\}$ is a relation on $A \times A \times A \times A$ with A as above.

3.4.2   *Arrow diagrams*

A relation between two sets can be visually represented with an arrow diagram. Suppose R is a relation from a set A to a set B. We can define the arrow diagram as follows:

1. Represent the elements of A as points in one region and the elements of B as points in another region.
2. For each $x \in A$ and $y \in B$, draw an arrow from x to y if and only if $(x,y) \in R$.

Example 3.4.3 Let $A = \{1,2,3\}$ and $B = \{1,3,5\}$. Define the relations S and T from A to B as follows: For all $(x,y) \in A \times B$,

- $(x,y) \in S$ iff $x < y$.
- $T = \{(2,1), (2,5)\}$.

Draw arrow diagrams for S and T.

Solution:

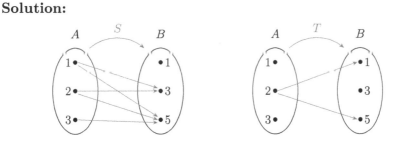

3.4.3 *Functions*

The word *function* indicates dependence of one set of values on another. The following shows a set of values, called input and the corresponding set of values, called output in the function named **Square**:

$$Square = \{(x,y) \mid y = x^2\}$$

Input $= \{-1,0,1,2,3\}$
Output $= \{0,1,4,9\}$
$Square =$
$\{(-1,1),(0,0),(1,1),(2,4),(3,9))$

Input	Output
-1	1
0	0
1	1
2	4
3	9

Definition 3.4.4 (Function) A *function* f from a set A to a set B, denoted

$$f : A \longrightarrow B$$

is a relation with the domain A and co-domain B that satisfies the following two properties:

1. For every element x in A, there is an element y in B such that $(x,y) \in f$. This is the "at least" property.
2. For all elements x in A and y and z in B, if $(x,y) \in f$ and $(x,z) \in f$, then $y = z$. This is the "at most" property.

A function is essentially a relation between two sets. To be a function, the relation must satisfy the above two properties. These two properties ensure that each element of A is associated with (we also say maps to) **exactly** one element of B. Note that the definition of a function consists of three things, namely, (1) the domain, (2) the co-domain, and (3) the association rule given by f.

Given a function $f : A \longrightarrow B$, and elements a in A and b in B, if $f(a) = b$, then b is called the *image* of a under f. We can similarly define the image of a subset of the domain under the function f. Let $C \subseteq A$, then the image of C under f, denoted $f(C)$ is defined by

$$f(C) = \{f(c) \,|\, c \in C\}.$$

In particular, the *range* of f is defined as the image of the domain of f. That is,

$$range(f) = f(A) = \{f(a) \,|\, a \in A\}.$$

Example 3.4.5 Let $A = \{2, 4, 6\}$ and $B = \{1, 3, 5\}$. Which of the following relations R, S, and T defined below are functions from A to B?

1. $R = \{(2,5),(4,1),(4,3),(6,5)\}$.
2. For all $(x,y) \in A \times B$, $(x,y) \in S$ iff $y = x + 1$.
3. T is defined by the following arrow diagram:

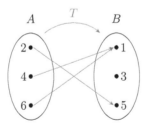

Solution:

1. *R satisfies the "at least" property. However, $4 \in A$ is associated with two elements in B: 1 and 3. So R does not satisfy the "at most" property. Therefore, R is not a function.*
2. *According to the association rule $y = x + 1$, S can be expressed as $S = \{(2,3),(4,5)\}$. Only two out of the three elements of A are associated with some elements of B. The element 6 of A is not associated with any element in B. Therefore, S is not a function.*

3. *According to the arrow diagram of T, 2 in A maps to 5 in B and both 4 and 6 in A map to 1 in B. So, both function properties are satisfied. Therefore, T is a function.*

Example 3.4.6 The following arrow diagram defines a function $f : A \longrightarrow B$:

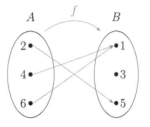

1. Determine the images of the following elements of $A : 2, 4, 6$.
2. Determine the images of the following subsets of $A : \{2, 4\}, \{2, 6\}$, $\{4, 6\}, \{2, 4, 6\}$.
3. What is the range of f?

Solution:

1. $f(2) = 5$, $f(4) = 1$, *and* $f(6) = 1$.
2. $f(\{2, 4\}) = \{1, 5\}$, $f(\{2, 6\}) = \{1, 5\}$, $f(\{4, 6\}) = \{1\}$, $f(\{2, 4, 6\}) = \{1, 5\}$.
3. *The range of f is* $\{1, 5\}$.

Exercises

Exercise 3.4.1 Let $E = \{2, 3, 4\}$ and $F = \{-2, -1, 0\}$. Define a relation T from E to F by

$$T = \left\{ (x, y) \in E \times F \,\middle|\, \frac{x - y}{3} \in \mathbb{Z} \right\}.$$

1. Is $(3, 0) \in T$? Is $(3, -1) \in T$? Is $(3, -2) \in T$?
2. Determine the domain and co-domain of T.
3. Draw an arrow diagram for T.

Exercise 3.4.2 Let $X = \{1, 3, 5\}$ and $Y = \{s, t, u, v\}$. Define the relation $S \subseteq X \times Y$ by the following arrow diagram:

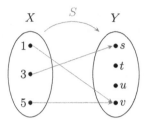

1. Determine the domain and co-domain of S.
2. Is S a function?
3. What is the range of S?
4. Represent S as a set of ordered pairs.

Exercise 3.4.3 Let $R \subseteq \mathbb{N} \times \mathbb{Z}$ be the relation defined by

$$R = \{(a, b) \mid a \text{ divides } b\}.$$

1. List 4 members of $\mathbb{N} \times \mathbb{Z}$ that are in R.
2. List 4 members of $\mathbb{N} \times \mathbb{Z}$ that are not in R.
3. Is R a function?

Chapter 4

Predicate Logic

As promised in Chapter 2, we wish to add to the expressive power of our language, propositional logic. We will do this by introducing variables, predicates, and functions. This will be done in the most general form for the sake of completeness. However, we start with a simple example to show you the idea and the expressive power of this new artificial language.

Let us consider an example language. Suppose we wish to design a language to talk about people, in particular, we single out two people called Peter and Jane. We would like to talk about love and hate relationships between people. We shall also need to refer to a person's father. Thus we choose for our language $\mathcal{L}$ to have the following symbols:

- **Constant symbols:** P, J.
- **Predicate symbols:** $H(x, y), L(x, y)$.
- **Function symbols:** $F(x)$.

The meaning we have in mind for these symbols is as follows: P stands for "Peter", J stands for "Jane". $H(x, y)$ means that person x hates person y, and $L(x, y)$ means that person x loves person y. Finally, $F(x)$ refers to x's father. Note that we have specified the meaning of the symbols of our language in an informal way. We shall keep this informal way of giving meaning to our symbols for a while longer. Later on, when we have more experience with the language we will show how to formally give meaning to the symbols of any such language.

Before we give examples of formulas in this language let us mention some symbols that come by default in any language. One such symbol is "=" which stands for equality relation between objects that the language intends to talk about. We certainly have access to all the logical connectives that we had before in propositional logic, but we get two new unary connectives called "quantifies". These are: $\forall$, called the *universal quantifier*, and $\exists$ which is called the *existential quantifier*. They read as you might expect: $\forall x$ is read as "for every x" and $\exists x$ is read as "there exists at least one x".

Let us now look at some formulas in this language we just introduced at the beginning of this chapter. For example,

- $\forall x\, L(x, P)$ says that everybody loves Peter.
- $\forall x \exists y H(x, y)$ says that everybody hates somebody.
- $L(P, J)$ says that Peter loves Jane.
- $L(P, F(J))$ says that Peter loves Jane's father.

So, we shall try to devise a language like this in its most general form and will call it the *predicate logic*, sometimes also called the *first-order logic*. You may wonder why it is called predicate logic. According to Oxford English dictionary, the word "predicate" means "the part of a sentence or clause containing a verb and stating something about the subject (e.g., went home in John went home)". And this is precisely what the language of predicate logic tries to do, that is, we denote the predicate "x went home" by $H(x)$ and use the symbol "J" in the language to refer to John and then we can say "John went home" in our formal language by simply stating $H(J)$. That is to say, this new language of predicate logic abstracts predicates into symbols of the language and then is able to express all such predicates about any given subjects, like John in the example above.

Let us look at another example. Suppose we want to express that "James is a student at Avalon University". We will use a constant symbol J that will stand for "James". We will then use a predicate symbol $S(x)$ to mean that x is a student at Avalon University. Then the statement above translates to $S(J)$ in our language.

And finally, let us look at an example from mathematics. Let $G(x, y)$ stand for "number x is greater than number y". Then, we can express "For every number there is a number greater than it" by the formula $\forall x \exists y G(y, x)$.

4.1 Syntax

We have the following alphabet that we will use to construct formulas:

- **Variables:** $x_1, x_2, x_3, \ldots, y_1, y_2, \ldots$
  There are countably many variables. In other words as many distinct variable names as you might need.
- **Constants:** $c_1, c_2, \ldots$
  There are countably many constant symbols. Again, as many as you might wish.
- **Predicate symbols:** We have predicate symbols of all arities (arity is the number of place holders, in other words the number of variables that the predicate can handle):
  $P_1(x_1), P_2(x_1), \ldots$
  $Q_1(x_1, x_2), Q_2(x_1, x_2), \ldots$
  $R_1(x_1, \ldots, x_k), R_2(x_1, \ldots, x_k), \ldots$
  Nullary (arity zero) predicates are just atomic propositions. We also have a special binary predicate symbol, denoted $=$.
- **Function symbols:** We have function symbols of all arities:
  $f_1(x_1), f_2(x_1), \ldots$
  $g_1(x_1, x_2), g_2(x_1, x_2), \ldots$
  $h_1(x_1, \ldots, x_k), h_2(x_1, \ldots, x_k), \ldots$
- **Logical connectives:** We have the logical connectives: $\neg, \wedge, \vee,$ $\longrightarrow, \leftrightarrow, \forall$ (Universal Quantifier), and $\exists$ (Existential Quantifier).
- Finally, we shall need left and right parentheses.

Before we can write correct (well-formed) formulas we need to introduce another concept, namely that of a term.

Definition 4.1.1 A *term* is defined as follows:

- All variables are terms.
- All constants are terms.
- If $t_1, t_2, \ldots, t_n$ are terms for $n \geq 1$ and f is an n-ary function symbol, then $f(t_1, t_2, \ldots, t_n)$ is a term.
- Nothing else is a term.

Note that, terms are supposed to refer to objects that the language is designed to talk about. For example, in the case of our language $\mathcal{L}$ above we wish to talk about Peter and Jane and their fathers, etc.

Here are some examples and non-examples of terms in our language $\mathcal{L}$ above.

Example 4.1.2

- The following are all terms: $F(F(x)), F(F(P)), J, x, y, P$.
  For example, $F(F(x))$ refers to x's grandfather. $F(F(P))$ refers to Peters grandfather, etc.
- The following are **not** terms: $L(x, y), L(P, F(P)), PP$.

It is important that you go over each of these and convince yourselves that they are not terms. For example, why is $L(x, y)$ not a term? Well, it is not a variable or a constant symbol and it is not built from variables or constant symbols using a function symbol. Note that L is a relation symbol. Also, you could check that it is not a term by asking what object it refers to. And you will realize that $L(x, y)$ talks about a relationship between objects, in this case x and y, and not objects themselves. So, it cannot be a term.

We shall next define the well-formed formulas:

Definition 4.1.3 Formulas are constructed from atomic formulas using logical connectives as follows:

- **Atomic formulas:** All 0-ary predicate symbols are atomic formulas. If $t_1, \ldots, t_n$ are terms and P is an n-ary predicate symbol with $n \geq 1$, then $P(t_1, t_2, \ldots, t_n)$ is an atomic formula. For example, $t_1 = t_2$ is an atomic formula.
- If A is a formula, then so is $\neg A$. If A and B are formulas, then so are $A \wedge B, A \vee B, A \longrightarrow B, A \leftrightarrow B$, and finally if A is a formula, then so are $\forall x(A)$ and $\exists x(A)$. We read $\forall x(A)$ as "for all x, A", and we read $\exists x(A)$ as "there exists x, A".

  Also $\forall x$ and $\exists x$ bind all the occurrences of x in the formula A. See below for a precise definition of free and bound occurrences of variables.

Note that formulas are there to express relationships between objects of study. So, they take on terms and say something about them. Such things may be true or false but that is not what we are concerned with now. We will talk about truth value evaluation of formulas in the next section.

Here are some examples and non-examples of formulas in our toy language $\mathcal{L}$ above:

Example 4.1.4

- The following are all formulas in $\mathcal{L}$:

$$L(F(P), P), H(F(P), F(F(J))), \forall x \forall y L(x, y), \text{ and}$$
$$\exists x \forall y H(y, x), L(P, J) \longrightarrow L(F(P), F(J)).$$

Please make sure you understand why. For example, $\forall x \forall y L(x, y)$ is a formula because it is formed starting with $L(x, y)$ which is a formula, then we apply $\forall y$ to a formula which yields a formula and then we apply $\forall x$ to that formula to get the final formula.

- None of the following expressions is a formula:

$$F(P), L(L(x, y), z), F(L(x, y)), L(H(x, y), L(y, x)).$$

For example, $F(L(x, y))$ is not a formula because even though it starts with a formula, namely $L(x, y)$, it applies a function symbol F to this formula which makes no sense. You can also see this by asking what kind of relationship this expression is trying to express. This seems to say "Father of x loves y" which makes no sense because "x loves y" is a not an object of our language, it is not a person.

Definition 4.1.5 An occurrence of a variable x in a formula is said to be a *bound* occurrence, if it is in the scope of a quantifier using the variable x. Otherwise, it is said to be a *free* occurrence.

Example 4.1.6

- The only occurrence of x in $\exists x(x + 3 = 7)$ is bound.
- In $\forall x(A(x) \wedge H(x, y))$, both occurrences of x are bound and the only occurrence of y is free.
- In $\forall x(H(x, x)) \wedge \exists y(H(x, y))$, the first two occurrences of x are bound, but the third is free, the y is also bound.

We shall continue our study of predicate logic by investigating semantic issues.

Exercises

Consider a first-order language $\mathcal{L}_1$ defined as follows:

- **Constant symbols:** O.
- **Predicate symbols:** $L(x, y)$.
- **Function symbols:** $S(x), P(x, y)$.

Exercise 4.1.1 Determine whether each of the following is a term in the language $\mathcal{L}_1$.

1. x,
2. $S(L(x, y))$,
3. $P(O, x)$,
4. $P(S(S(O)), O)$,
5. $L(P(x), y)$,
6. $P(L(x, y), y)$.

Exercise 4.1.2 Determine whether each of the following is a formula in the language $\mathcal{L}_1$.

1. $S(L(x, y))$,
2. $P(O, x)$,
3. $L(P(x, y), y)$,
4. $L(S(x), P(x, y))$,
5. $L(S(x), y)$,
6. $\forall x \forall y L(x, y)$,
7. $\exists x L(O, x)$.

4.2   Semantics

Let us consider again the language $\mathcal{L}$ where

- **Constant symbols:** P, J.
- **Predicate symbols:** $H(x, y), L(x, y)$.
- **Function symbols:** $F(x)$.

Also, suppose that we are given a formula $\forall x \exists y H(x, y)$ and asked to determine if this formula is true. If you think about this for a while, you notice that you cannot answer this question because you do not know what $H(x, y)$ means. Therefore, we need more information to be able to evaluate the truth value of any formula. This information comes in the form of a model and a variable assignment function.

We will first give the formal definitions and then follow with many examples.

Definition 4.2.1 A *model* M is a pair $M = (U, I)$ where U is a non-empty set called the *universe* and I is called an *interpretation* defined as follows:

- $I(c) \in U$, that is any constant symbol is interpreted as an element of U,
- $I(P) \in \{t, f\}$, for P a predicate symbol of arity 0,
- $I(P) \subseteq U^n$, for P a predicate symbol of arity $n \geq 1$,
- $I(f) : U^n \longrightarrow U$, for f a function symbol of arity $n \geq 1$.

In the definition above, U^n is the shorthand for the n-fold Cartesian product of U by itself.

A *variable assignment function* is a function from the set of variables V to the universe, that is, $\rho : V \longrightarrow U$.

Given a model and a variable assignment function we can determine the truth value of any formula. This will be formally defined below, however, we shall first look at some examples.

Example 4.2.2 Let us go back to our language $\mathcal{L}$ above, and suppose we are given the model $M = (\{a, b, c, d\}, I)$, where

- $I(P) = a, I(J) = b$,
- $I(H) = \{(a, a), (a, b), (b, a), (a, c), (a, d)\}$,
- $I(L) = \{(d, d), (d, c), (d, a), (b, a), (a, b)\}$, and
- $I(F)(i) = i$ for any $i \in \{a, b, c, d\}$.

Also assume that the variable assignment function is given by $\rho(x) = a, \rho(y) = b$.

Now, given this model M and variable assignment function ρ, let us determine the truth value of several different formulas in $\mathcal{L}$:

- Consider the formula $\forall x \exists y H(x, y)$. If this formula were true, then for each $x \in U$ there must be a $y \in U$ such that $(x, y) \in I(H)$, but for example, if you let $x = c$ then, there is no y such that $(c, y) \in I(H)$. So, the formula is false.
- Consider the formula $\forall x H(x, y)$. We know that the value for y is given as b (as $\rho(y) = b$), so if this formula were true, then for each value of $x \in U$ we must have $(x, b) \in I(H)$, however, if you let $x = d$, then $(d, b) \notin I(H)$, so the formula is false.
- Consider the formula $\exists x \forall y H(x, y)$, this formula is true because for $x = a$, and any value for $y \in U$, we have $(x, y) \in I(H)$.
- Consider the formula $L(F(P), F(J)) \longrightarrow \forall x L(x, J)$. The antecedent is true because $I(P) = a$, $I(J) = b$, and $I(F)(a) = a$ and $I(F)(b) = b$, and $(a, b) \in I(L)$. However, the consequent is false, because for example, $(d, b) \notin I(L)$. Therefore, the formula is false.
- Finally, $H(x, y)$ is true, because $(a, b) \in I(H)$.

Here is another example derived from the language $\mathcal{L}$ above.

Example 4.2.3 Consider the language $\mathcal{L}_1$ where we drop the function symbol F and keep everything else intact. Let us define a model M as follows.

- Suppose our intention is to talk about a group of friends. So, we take their names and form our universe U. Hence, $U = \{Peter, Jane, Bob, Alice, Dave\}$.
- $I(P) = $ Peter, and $I(J) = $ Jane. We do this because we wish to focus on these two individuals.
- L is meant to describe the love relationship amongst these friends and so to the best of our knowledge we shall list the pairs of people where the first person loves the second person. So, say we have $I(L) = \{(Peter, Jane), (Jane, Peter), (Bob, Alice)\}$. This means, for example, that Peter loves Jane, etc.
- We define $I(H) = \emptyset$.

As for the variable assignment function, we let $\rho(x) = Alice$, $\rho(y) = Bob$.

Given the language $\mathcal{L}_1$ and the model $M = (U, I)$ and variable assignment function ρ, determine the truth value of the following formula:

$$\forall x(L(x, P) \longrightarrow L(P, x)).$$

Solution. *Let us see what this formula is saying. The translation is "for every $x \in U$, if $L(x, P)$ is true, then $L(P, x)$ is also true". This simply says "if someone loves Peter, then Peter loves that person".*

The following table shows the truth value of the implication when x takes each of the values in the universe U.

x	$L(x, P)$	$L(P, x)$	$L(x, P) \longrightarrow L(P, x)$
Peter	False	False	True
Jane	True	True	True
Alice	False	False	True
Dave	False	False	True
Bob	False	False	True

The table above shows that the implication $L(x, P) \longrightarrow L(P, x)$ is true for all x in the universe. Therefore, the formula is true.

We shall formally define the meaning operation $[\![-]\!]$. The idea is that given a model M and a variable assignment function ρ, for any formula A, its meaning $[\![A]\!]_{M,\rho} \in \{t, f\}$.

Definition 4.2.4 Given a model $M = (U, I)$ and a variable assignment function $\rho : V \longrightarrow U$, the *meaning operation*, denoted $[\![-]\!]_{M,\rho}$ is defined as follows[1]:

- $[\![x]\!] = \rho(x)$, where x is a variable.
- $[\![c]\!] = I(c)$, where c is a constant symbol.
- $[\![f(t_1, \ldots, t_n)]\!] = I(f)([\![t_1]\!], \ldots, [\![t_n]\!])$, where f is a function symbol of arity $n \geq 1$.
- $[\![P]\!] = I(P)$, where P is a predicate symbol of arity 0.
- $[\![P(t_1, \ldots, t_n)]\!] = t$ iff $([\![t_1]\!], \ldots, [\![t_n]\!]) \in I(P)$, where P is a predicate symbol of arity $n \geq 1$.
- $[\![t_1 = t_2]\!] = t$ iff $[\![t_1]\!] = [\![t_2]\!]$.
- $[\![\neg A]\!] = t$ iff $[\![A]\!] = f$.
- $[\![A \wedge B]\!] = t$ iff $[\![A]\!] = t$ and $[\![B]\!] = t$.

---

[1] We shall drop the subscript M, ρ to avoid cumbersome expressions.

- $[\![A \vee B]\!] = t$ iff $[\![A]\!] = t$ or $[\![B]\!] = t$.
- $[\![A \longrightarrow B]\!] = f$ iff $[\![A]\!] = t$ and $[\![B]\!] = f$.
- $[\![A \leftrightarrow B]\!] = t$ iff $[\![A]\!] = [\![B]\!]$.
- $\big[\![\forall x A(x)]\!\big] = t$ iff $\big[\![A(x)]\!\big] = t$ for **all** values that x takes in U.
- $\big[\![\exists x A(x)]\!\big] = t$ iff $\big[\![A(x)]\!\big] = t$ for **some** value that x takes in U.

Note that, in general the meaning of a formula depends on the chosen model M, and the function ρ.

We say that a formula A is *true in a model* M, denoted $M \models A$ iff $[\![A]\!]_{M,\rho} = t$, for all ρ.

We say that a formula A is *universally valid*, denoted $\models A$ iff $M \models A$ for all models M. Universal validity plays the role of tautology in propositional logic.

Let us look at a few more examples.

Example 4.2.5 Let $\mathcal{L}$ be a language with a single binary predicate symbol $Q(x,y)$ and constant symbols $0, 1, 2, 3, 4, \ldots$. Let $M = (\mathbb{N}, I)$ where $\mathbb{N}$ is the set of natural numbers. All constant symbols are interpreted as usual natural numbers, i.e., $I(0) = 0, I(1) = 1$, $I(2) = 2$, etc., and $I(Q) = \{(m,n) \mid m = n+3\}$. What are the truth values of the following two formulas?

- $Q(1,2)$,
- $Q(3,0)$.

Solution:

- $Q(1,2)$ is false because $(1,2) \notin I(Q)$ as $1 = 2+3$ is not true.
- $Q(3,0)$ is true because $(3,0) \in I(Q)$ as $3 = 0+3$ is true.

Example 4.2.6 Let $\mathcal{L}_1$ be a language with a single ternary predicate symbol $R(x,y,z)$ and constant symbols $0, 1, 2, 3, 4, \ldots$. Let $M = (\mathbb{N}, I)$ where $\mathbb{N}$ is the set of natural numbers. All constant symbols are interpreted as usual natural numbers, i.e., $I(0) = 0, I(1) = 1$, etc., and $I(R) = \{(m,n,p) \mid m+n = p\}$. What is the truth value of each of the following formulas?

- $R(1,2,3)$,
- $R(0,0,1)$.

Solution:

- $R(1, 2, 3)$ *is true because* $(1, 2, 3) \in I(R)$ *as* $1 + 2 = 3$.
- $R(0, 0, 1)$ *is false because* $(0, 0, 1) \notin I(R)$ *as* $0 + 0 = 1$ *is false.*

Example 4.2.7 Let $\mathcal{L}_2$ be a language with a single binary predicate symbol $O(x, y)$ and three constant symbols J, P, and K. Let $M = (U, I)$ where U is the set of all people in a class, $I(J) = $ John, $I(P) = $ Peter, and $I(K) = $ Kim. $I(O) = \{(John, Peter), (Peter, Kim)\}$. $O(x, y)$ means x is older than y. What is the true value of each formula below?

1. $O(J, P)$,
2. $O(J, K)$,
3. $O(K, K)$.

Solution:

1. $O(J, P)$ *is true because* $(I(J), I(P)) = (John, Peter) \in I(O)$.
2. $O(J, K)$ *is false because* $(I(J), I(K)) = (John, Kim) \notin I(O)$.
3. $O(K, K)$ *is false because* $(I(K), I(K)) = (Kim, Kim) \notin I(O)$.

Here are a few more examples where we focus on quantifiers.

As we said in the previous section Predicate Logic has two types of quantification: universal quantification, and existential quantification. Universal quantification asserts a property or relationship that applies to every element or object in a given universe. For example, *all students at Avalon University must earn 2.5 or higher cumulative GPA to earn a Bachelor's degree* is a universal statement.

Existential quantification asserts the existence of at least one element or object in a universe that satisfies a particular property or condition. For example, *some students at Avalon University graduate with honors* is an existential statement. Note that there may be more elements of the universe that satisfy the property, or even all may satisfy the property but all we need for an existentially quantified statement to be true is to have at least one example that satisfies the desired property.

Example 4.2.8 Let $\mathcal{L}$ be a language with a single unary predicate symbol $P(x)$. Consider the model $M = (U, I)$ where $U = \mathbb{R}$

(the set of real numbers) and $I(P) = \{x \in \mathbb{R} \,|\, x + 1 > x\}$, and $\rho(x) = \sqrt{2}$ and $\rho(y) = 1/3$. What is the truth value of $\forall x P(x)$?

Solution: *First note that we do not need the variable assignment function ρ to determine the truth value of this formula because there are no free variables in this formula. The formula $\forall x P(x)$ is true. In order to show that we need to show that for any real number r, $r \in I(P)$. That is, we need to show that for any real number r, $r + 1 > r$, which is obviously true.*

Example 4.2.9 Given the same language and model as above determine the truth value of the formula $x < 2$.

Solution: *In order to evaluate the truth value of this formula we need the function ρ because the variable x occurs freely in this formula. We know that $\rho(x) = \sqrt{2}$ and that $\sqrt{2} < 2$. Therefore, the formula is true in this model and environment.*

Example 4.2.10 Given the same language and model as above determine the truth value of the formula $\exists x P(x)$?

Solution: *The formula $\exists x P(x)$ is true. To see this, note that all we need to show is that there is a real number r such that $r + 1 > r$. Let us pick $r = 2$, clearly $2 + 1 > 2$, and so the formula above is true.*

Very often textbooks in discrete mathematics discuss predicate logic in a very different way. The language is never explicitly given, except for its symbols for predicates and functions. All constant, predicate, and function symbols come interpreted. So, when you see $x > y$ in a math book it means that your language has a binary predicate ">" which is already interpreted as the greater than symbol over a given set (domain). We have been translating such problems from discrete math to our way of dealing with logical formulas in the examples above. We think that the method pursued here in this book is more appropriate for an approach to logic that may go beyond mathematical applications. However, we shall give you some examples below in the style of the discrete math books.

Example 4.2.11 What is the truth value of $\forall x Q(x)$, where $Q(x)$ stand for $x > 2$ and the domain consists of all real numbers?

Solution: *The formula* $\forall x Q(x)$ *is false. To justify it, we need to show* $Q(x)$ *is false for some real number because the quantifier is the universal quantifier. Here is a counterexample: Let* $x = 1$, *then* $Q(x)$ *is false because* $1 > 2$ *is false.*

Example 4.2.12 Let $Q(x)$ denote the statement "$x = x+1$". What is the truth value of $\exists x Q(x)$ where the domain consists of all real numbers?

Solution: *The formula* $\exists x Q(x)$ *is false. To justify it, we need to show* $Q(x)$ *is false for all real numbers because the quantifier is the existential quantifier.*

$x = x + 1$ *is always false because otherwise, if it were true for some* x, *then we subtract* x *from both sides and get* $0 = 1$! *Therefore,* $x = x + 1$ *is false for all real numbers.*

Example 4.2.13 What is the truth value of $\exists x T(x)$, where $T(x)$ is the statement "$x^2 > 10$" and the domain of x consists of positive integers not exceeding 4?

Solution: *The formula* $\exists x T(x)$ *is true. Given that the quantifier is an existential quantifier, we need to show* $T(x)$ *is true for some positive integer less than or equal to 4. Indeed, if we pick* $x = 4$, *then as* $4^2 > 10$ *is true we have shown that the formula is true.*

So far, we have seen formulas with a single quantifier. However, as we have seen we can have any number of predicates in a formula with some quantifiers in the scope of other quantifiers.

Definition 4.2.14 (Nested quantifiers) Two quantifiers are nested if one is within the scope of the other. For example, $\forall x \exists y (x + y) = 0$. Here the existential quantifier is in the scope of the universal one.

Example 4.2.15 Translate the following formulas into English. The domain of all variables is the set of all real numbers.

1. $\forall x \forall y (x + y = y + x)$,
2. $\forall x \exists y (x + y = 0)$,
3. $\forall x \forall y ((x > 0) \wedge (y < 0) \longrightarrow (xy < 0))$.

Solution:

1. *For any two real numbers x and y, the sum of x and y is equal to the sum of y and x.*
2. *For any real number x, there exists some real number y such that the sum of x and y is equal to 0.*
3. *For any real numbers x and y, if x is positive and y is negative, then the product of x and y is negative.*

Example 4.2.16 Determine the truth value of the formulas in the previous example.

Solution:

1. *This formula is true because we know that the addition operation on the set of real numbers is commutative.*
2. *This formula is also true. Given a real number x, let $y = -x$ which is clearly a real number and $x + y = x + (-x) = 0$.*
3. *This formula is also true, because the product of a positive and a negative real number is a negative real number.*

We say that A is *logically equivalent* to B, denoted $A \equiv B$ iff $A \leftrightarrow B$ is universally valid. Here are some examples of logical equivalences:

- $\exists x \exists y A(x, y) \equiv \exists y \exists x A(x, y)$,
- $\forall x \forall y A(x, y) \equiv \forall y \forall x A(x, y)$.
- De Morgan Laws:
  - $\neg(\forall x A(x)) \equiv \exists x \neg(A(x))$,
  - $\neg(\exists x A(x)) \equiv \forall x \neg(A(x))$.

As you can see in the first two logical equivalences, the order of quantification does not matter provided all the quantifiers are of the same kind. However, if the quantifiers are of different kinds, then the order does matter. We will see examples later below.

Here are some examples of the application of De Morgan laws.

Example 4.2.17 Simplify the following formulas such that $\neg$ occurs in front of the atomic formulas only.

- $\neg(\forall x(\exists y P(x, y, z) \wedge \forall z P(x, y, z)))$,
- $\neg(\exists x(P(y, z) \vee (\forall y(\neg Q(y, x) \wedge P(y, z)))))$.

Solution:

$$\neg(\forall x(\exists y P(x,y,z) \wedge \forall z P(x,y,z)))$$
$$= \exists x \neg(\exists y P(x,y,z) \wedge \forall z P(x,y,z))$$
$$= \exists x(\neg(\exists y P(x,y,z)) \vee \neg(\forall z P(x,y,z)))$$
$$= \exists x(\forall y \neg P(x,y,z) \vee \exists z \neg P(x,y,z)),$$

$$\neg(\exists x(P(y,z) \vee (\forall y(\neg Q(y,x) \wedge P(y,z)))))$$
$$= \forall x \neg(P(y,z) \vee (\forall y(\neg Q(y,x) \wedge P(y,z))))$$
$$= \forall x(\neg P(y,z) \wedge \neg(\forall y(\neg Q(y,x) \wedge P(y,z))))$$
$$= \forall x(\neg P(y,z) \wedge (\exists y \neg(\neg Q(y,x) \wedge P(y,z))))$$
$$= \forall x(\neg P(y,z) \wedge (\exists y(Q(y,x) \vee \neg P(y,z)))).$$

An argument

$$A_1$$
$$A_2$$
$$\vdots$$
$$\frac{A_n}{C}$$

is said to be *valid in a model M and an environment* ρ iff

$$[\![(A_1 \wedge \cdots \wedge A_n) \longrightarrow C]\!]_{M,\rho} = t.$$

It is said to be *valid in a model M*, iff

$$M \models (A_1 \wedge \cdots \wedge A_n) \longrightarrow C,$$

and to be *universally valid* iff

$$\models (A_1 \wedge \cdots \wedge A_n) \longrightarrow C.$$

Let us look at some examples.

Example 4.2.18 Consider the language $\mathcal{L}$ defined as follows:

- **Constant symbols:** *alma*.
- **Predicate symbols:** *loves(x, y)*.
- **Function symbols:** *next(x)*.

Suppose we are given the model $M = (\{a, b, c\}, I)$ where

- $I(alma) = a$,
- $I(loves) = \{(a, a), (a, b), (a, c), (b, a)\}$,
- $I(next)(a) = b, I(next)(b) = a, I(next)(c) = c$,

and the function ρ with $\rho(x) = a, , \rho(y) = b, \rho(z) = b$.

- $[\![\forall x \exists y\, loves(x, y)]\!] = f$, *because if we let $x = c$, then there is no value for y in $\{a, b, c\}$ such that $(c, y) \in I(loves)$.*
- $[\![\exists z\, loves(alma, z)]\!] = t$, *because if we let $z = a$, we observe that $([\![alma]\!], a) = (a, a) \in I(loves)$. Similarly $[\![\forall z\, loves(alma, z)]\!] = t$, because $(a, a), (a, b), (a, c)$ are all in $I(loves)$.*
- $[\![\exists x\, loves(next(x), alma)]\!] = t$, *because if we let $x = a$ then $([\![next(x)]\!], [\![alma]\!]) = (I(next)(a), [\![alma]\!]) = (b, a) \in I(loves)$.*
- $[\![loves(x, y)]\!] = t$, *because $([\![x]\!], [\![y]\!]) = (\rho(x), \rho(y)) = (a, b) \in I(loves)$.*

Example 4.2.19 Now consider the language $\mathcal{L}_a$ where

- **Constant symbols:** 0.
- **Predicate symbols:** $L(x, y)$.
- **Function symbols:** $S(x), P(x, y)$.

Suppose the model given is $(\mathbb{N}, I)$ where $I(0) = 0, I(L) = \{(m, n) \mid m \leq n\}, I(S) : \mathbb{N} \longrightarrow \mathbb{N}, I(S)(n) = n + 1$ and $I(P) : \mathbb{N} \times \mathbb{N} \longrightarrow \mathbb{N}, I(P)(m, n) = m + n$.

- $[\![\forall x L(0, x)]\!] = t$, *because 0 is interpreted as number 0 and this formula says that 0 is less than or equal to every natural number.*
- $[\![\forall x \exists y L(x, y)]\!] = t$, *because this formula says that for every natural number that we might pick there is always one that is greater than or equal to it.*
- $[\![\exists x \forall y L(y, x)]\!] = f$, *because this formula says that there is natural number that is greater than or equal to all natural numbers and we know that there is no such number.*

- $\llbracket \forall x (L(x, 0) \longrightarrow L(P(x, S(0)), 0)) \rrbracket = f$. *The only value for x in $\mathbb{N}$ making the antecedent true is $x = 0$, however then consequent is false. Note that $\llbracket P(x, S(0)) \rrbracket = I(P)(0, 1) = 0 + 1 = 1$ and $1 \not\leq 0$.*

Here are a few examples of universally valid arguments.

Example 4.2.20 Suppose we have a language $\mathcal{L}$ with a constant symbol Sam and two predicate symbols $H(x)$ and $M(x)$. Show that the argument below in $\mathcal{L}$ is universally valid.

$$\frac{\begin{array}{l} \forall x (H(x) \longrightarrow M(x)) \\ H(Sam) \end{array}}{M(Sam)} .$$

Solution: *Let $M = (U, I)$ be any model and $\rho : V \longrightarrow U$ be any variable assignment function. Also suppose the first hypothesis is true in this model, then it means that for any value $u \in U$ that x assumes, if $H(u)$ is true, then so is $M(u)$. But this means that if $u \in I(H)$ then $u \in I(M)$, which is simply saying that $I(H) \subseteq I(M)$. Now suppose the second hypothesis is true, then it means that $I(Sam) \in I(H)$. But, as $I(H)$ is a subset of $I(M)$, so $I(Sam) \in I(M)$ and hence the conclusion is true. Note that we made no assumptions on what the set U should be or what I is, so the argument is universally valid, that is, it is valid no matter what model we choose to work with. In terms of sets, this is just to say that if a set A is a subset of a set B, then any element a of A is also in B.*

Here is another example of a universally valid argument.

Example 4.2.21 Suppose we have a language $\mathcal{L}$ with three predicate symbols $A(x), B(x)$, and $C(x)$. Show that the argument below in $\mathcal{L}$ is universally valid.

$$\frac{\begin{array}{l} \exists x (A(x) \wedge B(x)) \\ \forall y ((A(y) \wedge B(y)) \longrightarrow C(y)) \end{array}}{\exists z C(z)} .$$

Solution: *Let $M = (U, I)$ be any model and $\rho : V \longrightarrow U$ be any variable assignment function. Also suppose the first hypothesis is true in this model, then it means that there is a value $u \in U$ that x assumes, and is such that $A(u)$ and $B(u)$ are both true, that is, there* is a $u \in I(A) \cap I(B)$. *Now suppose the second hypothesis is true, then it means that for any value $t \in U$ that y assumes, the formula is true, that is if $A(t) \wedge B(t)$ is true then so is $C(t)$, this means that for any $t \in U$, if $t \in I(A) \cap I(B)$, then $t \in I(C)$, which is just to say that $I(A) \cap I(B) \subseteq I(C)$. Now, as we know that there is a $u \in I(A) \cap I(B)$, then $u \in I(C)$. Therefore, the conclusion is true as well. Note that we made no assumptions on what the set U should be or what I is, so the argument is universally valid, that is, it is valid no matter what model we choose to work with.*

Here are a few examples on defining models.

Example 4.2.22 Consider the formula ϕ given by $\exists x \, (S(x) \wedge \forall y \, (S(y) \longrightarrow x = y))$. Define a model $M = (\{a, b, c, d\}, I)$ such that ϕ is valid in M. Also define a model $M' = (\{a, b, c, d\}, I')$ where ϕ is invalid.

Solution: *Let $I(S) = \{a\}$ and check that ϕ is valid in M. Then, let $I'(S) = \{a, b\}$ and check that ϕ is invalid in M'.*

Note that the formula can be translated into English saying that "There is exactly one thing satisfying the property S". For more on translation see the next section.

Example 4.2.23 Consider the formula ϕ given by $\forall x \, \exists y \, P(x, y) \longrightarrow \exists y \, \forall x \, P(x, y)$. Define a model $M = (\{a, b, c, d\}, I)$ such that ϕ is valid in M. Also define a model $M' = (\{a, b, c, d\}, I')$ where ϕ is invalid.

Solution: *In order to make ϕ valid (true) in M, we can try to make either the antecedent false or the consequent true. Let's try to make the antecedent false: let $I(P) = \{(a, a)\}$.*

To make ϕ false, we have to make the antecedent true and the consequent false: for example, $I'(P) = \{(a, a), (b, b), (c, c), (d, d)\}$.

Make sure you understand how this works.

Example 4.2.24 Consider the formula ϕ given by $\forall x \, \forall y \, \forall z \, ((R(x, y) \wedge R(y, z)) \longrightarrow R(x, z))$. Define a model $M = (\{a, b, c, d\}, I)$ such that ϕ is valid in M. Also define a model $M' = (\{a, b, c, d\}, I')$ where ϕ is invalid.

Solution: *Let $I(R) = \{(a,b),(b,c),(a,c)\}$, you can easily check that ϕ is true in this model.*

To make ϕ false, we have to make the antecedent true and the consequent false: for example, $I'(R) = \{(a,a),(a,b),(b,c),(d,d)\}$. Here, although (a,b) and (b,c) are in $I'(R)$, $(a,c) \notin I'(R)$.

Make sure you understand how this works.

Exercises

Exercise 4.2.1 Let $\mathcal{L}$ be a first order language with one unary predicate symbol S, one binary predicate symbol T, and one binary function symbol P. Consider the formulas

- $\phi_1 = \forall x \forall z \exists y[(x \neq z) \longrightarrow \neg(T(x,y) \wedge T(z,y))]$,
- $\phi_2 = \forall x \forall y \exists z[z = P(x, P(x,y))]$,
- $\phi_3 = \forall x[\forall z(\exists y(T(x,y) \longrightarrow T(z,y)) \wedge S(x)) \vee T(y,z)]$.

Simplify the formulas $\neg\phi_1$ and $\neg\phi_2$, transforming them such that $\neg$ is in front of atomic formulas only.

Exercise 4.2.2 Determine the free and bound occurrences of all variables in ϕ_1, ϕ_2 and ϕ_3 in the exercise above.

Exercise 4.2.3 Consider a first order language $\mathcal{L}$ that consists of two unary predicate symbols P and Q and one binary predicate symbol S. Also consider the following formulas in this language:

- $\phi_1 \equiv \forall y \exists x S(x,y) \wedge \exists x \forall y S(x,y)$,
- $\phi_2 \equiv \forall x(P(x) \vee Q(x)) \longrightarrow \forall x P(x) \vee \forall x Q(x)$.

Recall that a model M is a pair $M = (U, I)$, let $U = \{a,b,c,d\}$. Find one model $M_1 = (U, I_1)$ making ϕ_1 true and another model $M_2 = (U, I_2)$ making ϕ_2 false.

Exercise 4.2.4 Consider a first order language $\mathcal{L}$ that consists of one unary predicate symbol Q. Let ϕ be the formula:

$$\forall x \forall y \forall z[(Q(x) \wedge Q(y) \wedge Q(z)) \longrightarrow ((z = y) \vee (z = x) \vee (x = y))].$$

(i) Let $U = \{a,b,c,d\}$ and $I(Q) = \{a,b\}$. Is ϕ valid in this model?
(ii) Let $U' = \{a,b,c\}$ and $I'(Q) = \{a\}$. Is ϕ valid in this model?

Exercise 4.2.5 Show that the following argument is universally valid.

$$\frac{\forall x M(x)}{\exists y M(y)}.$$

4.3   Translation to Predicate Logic

In this section, we shall study translation from English to predicate logic. This is of course more involved than is the case for propositional logic.

We shall consider 6 general forms or patterns and discuss their translations. These will be very helpful in translating more complicated sentences into the language of predicate logic. For more details on stylistic variations see Appendix B.

- All A's are B's: $\forall x\,(A(x) \longrightarrow B(x))$.
- Some A's are B's: $\exists x\,(A(x) \wedge B(x))$.
- Some A's are not B's: $\exists x\,(A(x) \wedge \neg B(x))$.
- Not all A's are B's: $\exists x\,(A(x) \wedge \neg B(x))$.
- Only A's are B's: $\forall x\,(B(x) \longrightarrow A(x))$.
- No A's are B's: $\forall x\,(A(x) \longrightarrow \neg B(x))$.

Here are some examples. Note that we will assume (unless explicitly stated otherwise) that the universe of discourse for all English translations is the set of all things. So, in each case we have to specify further predicates about things, for example, whether they are people, stones, etc. Also in each case the interpretation of each predicate letter that we use will be clear from the context, in some cases we make this explicit if there is a possibility for ambiguity.

1. All cats will purr if their ears are rubbed.
   $\forall x\,(C(x) \longrightarrow (R(x) \longrightarrow P(x)))$.

2. Only persons over 21 will be admitted.
   $\forall x\,(A(x) \longrightarrow (P(x) \wedge O(x)))$.

3. Every student is younger than some instructor.
   $\forall x\,(S(x) \longrightarrow \exists y\,(I(y) \wedge Y(x,y)))$.

4. Not all birds can fly.
   $\exists x\,(B(x) \wedge \neg F(x)).$

5. Every child is younger than his mother.
   $\forall x \,\forall y\,((C(x) \wedge M(y,x) \longrightarrow Y(x,y)).$ Here $M(x,y)$ means that x
   is y's mother.

6. Some state is larger than Indiana.
   $\exists x\,(S(x) \wedge L(x,I)).$ Here I is the constant symbol that stands
   for Indiana.

7. Every number is larger than zero.
   $\forall x\,(N(x) \longrightarrow L(x,0)).$

8. Any healthy baby is pleased if people sing to him or show him
   bright objects.
   $\forall x\,((B(x) \wedge H(x)) \longrightarrow ((O(x) \wedge S(x)) \longrightarrow P(x))).$

9. No valuable diamonds are cracked or cloudy.
   $\forall x\,((D(x) \wedge V(x)) \longrightarrow \neg(C(x) \vee L(x))).$

10. Everything is caused by something.
    $\forall x\,\exists y\,C(y,x).$ Here $C(u,v)$ means u causes v.

11. There is something that causes everything.
    $\exists x\,\forall y\,C(x,y).$

We can also express some quantitative expressions in predicate
logic.

1. There is at least one student.
   $\exists x\,S(x).$

2. There are at least two students.
   $\exists x\,\exists y\,(S(x) \wedge S(y) \wedge (x \neq y)).$

3. There are at least three students.
   $\exists x\,\exists y\,\exists z\,(S(x) \wedge S(y) \wedge S(z) \wedge x \neq y \wedge x \neq z \wedge y \neq z).$

4. There is at most one student.
   $\forall x\,\forall y\,((S(x) \wedge S(y)) \longrightarrow x = y).$

5. There are at most two students.
   $\forall x\,\forall y\,\forall z\,((S(x) \wedge S(y) \wedge S(z)) \longrightarrow (x = y \vee x = z \vee y = z)).$

6. There is exactly one student.
   $\exists x\,(S(x) \wedge \forall y\,(S(y) \longrightarrow x = y)).$

We shall continue with more translation examples.

1. Everyone gave everyone something.
   $\forall x \, \forall z \, \exists y \, (G(x, y, z) \wedge P(x) \wedge P(z))$. Here $G(x, y, z)$ means x gave y to z and $P(x)$ means x is a person.

2. Everyone has a specific thing she gave to everyone.
   $\forall x \, \exists y \, \forall z \, (G(x, y, z) \wedge P(x) \wedge P(z))$.

3. There is some one thing and everyone gave it to everyone.
   $\exists y \, \forall x \, \forall z \, (G(x, y, z) \wedge P(x) \wedge P(z))$.

4. Len is the only person smarter than John.
   $P(L) \wedge P(J) \wedge S(L, J) \wedge \forall x \, ((P(x) \wedge x \neq J \wedge S(x, J)) \longrightarrow x = L)$.
   Here L stands for Len and J stands for John, and $S(x, y)$ means x is smarter than y.

5. Len is the tallest person in Bloomington.
   $P(L) \wedge B(L) \wedge \forall x \, ((P(x) \wedge B(x) \wedge x \neq L) \longrightarrow T(L, x))$. Here $B(x)$ means x is in Bloomington and $T(x, y)$ means x is taller than y.

6. Here is an example of translation from logic into English. Suppose the universe of discourse is the set of students in the computer club. $T(x, y)$ means "x called y", and $G(x, y)$ means "x gave y a gift". Then, $\forall x \, \forall y \, (x \neq y \longrightarrow (T(x, y) \wedge G(x, y)))$ is translated as

   > Every student in the computer club called and gave a gift to every other student in the club.

Exercises

Exercise 4.3.1 Let $L(x, y)$ stand for "x has sent y a letter", and $T(x, y)$ stand for "x has texted y". Also let N stand for Nida, and K stand for Keno.

Assume that the universe of discourse is the set of all students in your class.

Translate the following sentences to predicate logic:

a. No one in your class has texted Nida.
b. There is a student in your class who has not received a letter from Keno.
c. There are at least two students in your class such that one student has sent the other a letter and the second student has texted the first student.

Exercise 4.3.2 Let $L(x)$ mean "x has taken a logic course", and $H(x, y)$ mean "x has helped y".

Assume that the universe of discourse is the set of all students in your class.

Translate the following sentences to predicate logic:

a. No one in your class has taken a logic course.
b. There is a student in your class who has taken a logic course.
c. Every student in your class who has taken a logic course has helped every student in your class.

Exercise 4.3.3 Using the meaning for the predicate symbols as in the exercise immediately above, translate the following formulas into fluent English.

a. $\exists x \forall y (y \neq x \longrightarrow H(x, y))$,
b. $\forall x \exists y \, H(x, y)$,
c. $\exists x \, \exists y \forall z [(x \neq y) \wedge (H(x, z) \leftrightarrow H(y, z))]$.

4.4 Formal and Informal Proofs

We introduce the natural deduction proof system for predicate logic. The rules, as always come in pairs, an introduction rule and an elimination rule for each connective. The rules for all connectives, except the quantifiers are the same as in propositional logic.

∗ *Universal Quantifier:*

• **∀-Introduction**

$$
\begin{array}{ll}
1 & \vdots \\
2 & u \;\; \vdots \\
3 & \quad\;\; A(u) \\
4 & \forall x\, A(x) \qquad \forall\text{I, 2–3.}
\end{array}
$$

Here the variable u should not occur anywhere outside the subproof labelled by u. This assumption corresponds to the fact that one chooses an arbitrary u to prove $A(u)$. It is only after one achieves this, that one can claim to have proven $A(x)$ for all x.

- **∀-Elimination**

$$
\begin{array}{l|l}
1 & \vdots \\
2 & \forall x A(x) \\
3 & \vdots \\
4 & A(t) \qquad\qquad \forall\text{E, 2.}
\end{array}
$$

Here t is any term and $A(t)$ stands for the formula obtained from A by substituting the term t for all **free** occurrences of x in A. This rule simply says that if $A(x)$ is true for all x, then in particular it is true for any term t whatsoever.

* *Existential Quantifier:*

- **∃-Introduction**

$$
\begin{array}{l|l}
1 & \vdots \\
2 & A(t) \\
3 & \vdots \\
4 & \exists x A(x) \qquad\qquad \exists\text{I, 2.}
\end{array}
$$

Here t is any term. This rule simply says that if you have proven $A(x)$ for some term t, then you have proven that for some x, $A(x)$ is the case.

- **∃-Elimination**

$$
\begin{array}{l|l}
1 & \exists x A(x) \\
2 & \vdots \\
3 & u \;\big|\; A(u) \\
4 & \quad\;\; \vdots \\
5 & \quad\;\; B \\
6 & B \qquad\qquad \exists\text{E, 1, 3–5.}
\end{array}
$$

Here u is any variable that does not occur anywhere outside the subproof labelled by u and does not occur in B. This rule simply says that to prove a formula B from $\exists x A(x)$, just assume that $A(x)$ is the case for u substituted for all free occurrences of x in A, that is assume $A(u)$ and then try to prove B.

Here are some examples. We shall give a few examples of formal proofs below, but we will also give informal proofs.

Example 4.4.1 Give a formal and an informal proof for

$$\forall x(P(x) \longrightarrow Q(x)), P(I) \vdash Q(I).$$

Here I is a constant symbol.

Solution:

1	$\forall x(P(x) \longrightarrow Q(x))$	
2	$P(I)$	
3	$P(I) \longrightarrow Q(I)$	$\forall E,\ 1$
4	$Q(I)$	$\longrightarrow -E,\ 2,\ 3.$

Informally, we wish to prove $Q(I)$, but we know that for any x, if we have $P(x)$ then we get $Q(x)$. But we do have $P(I)$, so we get $Q(I)$.

Example 4.4.2 Give a formal and an informal proof for

$$\forall x(P(x) \longrightarrow Q(x)), P(I) \vdash \exists x Q(x).$$

Here, I is a constant symbol.

Solution:

1	$\forall x(P(x) \longrightarrow Q(x))$	
2	$P(I)$	
3	$P(I) \longrightarrow Q(I)$	$\forall E,\ 1$
4	$Q(I)$	$\longrightarrow -E,\ 2,\ 3$
5	$\exists x Q(x)$	$\exists I,\ 4.$

Informally, we wish to prove $Q(x)$ for some x, but we know that for any x, if we have $P(x)$ then we get $Q(x)$. We do have $P(I)$, so we get $Q(I)$. Hence, we have proven $Q(x)$ for some x, namely for $x = I$.

Example 4.4.3 Give a formal and an informal proof for

$$\exists x(P(x) \longrightarrow \forall x Q(x)) \vdash \forall x P(x) \longrightarrow \forall x Q(x).$$

Solution:

1	$\exists x(P(x) \longrightarrow \forall x Q(x))$	
2	$\quad \forall x P(x)$	
3	$\quad \exists x(P(x) \longrightarrow \forall x Q(x))$	*Reit.*, 1
4	$\quad u \mid P(u) \longrightarrow \forall x Q(x)$	
5	$\quad\quad \forall x P(x)$	*Reit.*, 2
6	$\quad\quad P(u)$	$\forall E$, 5
7	$\quad\quad \forall x Q(x)$	$\longrightarrow -E$, 4, 6
8	$\quad \forall x Q(x)$	$\exists E$, 3, 4–7
9	$\forall x P(x) \longrightarrow \forall x Q(x)$	$\longrightarrow -I$, 2–8.

Informally, we wish to prove that if everything is P, then everything is Q. Suppose everything is P. We know from hypothesis (1), that there is something, say u, such that if $P(u)$ then everything is Q. But as everything is P, in particular $P(u)$, so we have that everything is Q.

Example 4.4.4 Give a formal and an informal proof for

$$\forall x(P(x) \longrightarrow Q(x)) \vdash \forall x P(x) \longrightarrow \forall x Q(x).$$

Solution:

$$
\begin{array}{lll}
1 & \forall x(P(x) \longrightarrow Q(x)) & \\
\\
2 & \quad \forall x P(x) & \\
\\
3 & \quad\quad u \mid \forall x(P(x) \longrightarrow Q(x)) & \textit{Reit.}, 1 \\
\\
4 & \quad\quad\quad P(u) \longrightarrow Q(u) & \forall E,\ 3 \\
\\
5 & \quad\quad\quad \forall x P(x) & \textit{Reit.}, 2 \\
\\
6 & \quad\quad\quad P(u) & \forall E,\ 5 \\
\\
7 & \quad\quad\quad Q(u) & \longrightarrow -E,\ 4,\ 6 \\
\\
8 & \quad\quad \forall x Q(x) & \forall I,\ 3\text{--}7 \\
\\
9 & \quad \forall x P(x) \longrightarrow \forall x Q(x) & \longrightarrow -I,\ 2\text{--}8.
\end{array}
$$

Informally, we wish to prove that, if everything is P, then everything is Q. Suppose everything is P, from the hypothesis (1) we know that if anything is P, then it is Q. In particular as everything is P, so is u, and so $Q(u)$. But we have picked this u completely arbitrarily, thus we have proven that everything is Q.

We will look at two more examples of proofs.

Example 4.4.5 Give an informal proof for the argument below. Here, $LO(x, y), L(x), C(x)$ are predicate symbols and K is a constant symbol.

$$
\begin{array}{l}
1\ \forall x\,(C(x) \longrightarrow L(x)) \\
2\ \forall x\,(L(x) \longrightarrow LO(x, K)) \\
3\ \exists x\,C(x) \\
\hline
\exists x\,(L(x) \wedge LO(x, K))
\end{array}\ .
$$

Solution: *From (3), we have a t such that $C(t)$ holds, but from (1), we have that $L(t)$ holds and from (2) we see that $LO(t, K)$ holds too. Therefore, we have found something, namely t such that $L(t)$ and $LO(t, K)$ both hold and hence the conclusion follows.*

Example 4.4.6 Give an informal proof for the argument below. Here $L(x), C(x)$, and $D(x)$ are predicate symbols.

1 $\forall y\, (C(y) \vee D(y))$
2 $\forall x\, (C(x) \longrightarrow L(x))$
3 $\exists x\, \neg L(x)$
———————————
$\qquad \exists x\, D(x)$.

Solution: *From (3), we have a t such that $\neg L(t)$ holds, but from (2), using contraposition we have that $\neg C(t)$ holds and finally from (1) we see that $D(t)$ holds. Therefore, we have found something, namely t such that $D(t)$ holds and hence the conclusion follows.*

Exercises

Exercise 4.4.1 Give a formal proof for the sequent,

$$\frac{\forall y\, A(y)}{\exists x\, A(x)} .$$

Exercise 4.4.2 Give a formal proof for the sequent,

$$\frac{\forall y\, A(y)}{\exists x\, (A(x) \vee B(x))} .$$

Exercise 4.4.3 Give an informal proof for the sequent,

$$\frac{\exists y\, A(y) \quad \forall x\, (A(x) \longrightarrow (B(x) \wedge C(x)))}{\exists z\, B(z)} .$$

Exercise 4.4.4 Give an informal proof for the sequent,

$$\frac{\forall x\, (A(x) \longrightarrow B(x)) \quad \forall x\, (B(x) \longrightarrow C(x))}{\exists x\, (A(x) \longrightarrow C(x))} .$$

Chapter 5

Mathematical Induction

The principle of mathematical induction is a proof method that is used to prove universal statements that specify some properties for the set of natural numbers. For example, suppose we wish to prove that

$$1 + 2 + \cdots + n = \frac{n(n+1)}{2},$$

for any natural number $n \geq 1$. As a matter of fact, mathematical induction works in more general sets than just $\mathbb{N}$, but for our purposes we shall just work with $\mathbb{N}$. It is called mathematical induction to distinguish it from physical induction. To motivate the main idea behind induction, suppose the members of the set $\mathbb{N}$ are arranged on a line with 0 sitting first and then 1 next to 0 etc. We say that $i + 1$ is the right member of i. Now suppose you wish to spread a rumor to the members of $\mathbb{N}$, in other words, you want all the elements of $\mathbb{N}$ to know about something. The principle of mathematical induction says that it suffices to let 0 know about this rumor and to ensure that as soon as any i knows about it, she tells her right neighbor about it. One can easily see that given time everybody will know about this rumor following this protocol.

We shall write this formally: suppose $P(x)$ is a predicate about natural numbers. And that we wish to show that $\forall x\, P(x)$ is true.

5.1 Principle of Mathematical Induction

$$(P(0) \wedge \forall k\, (P(k) \longrightarrow P(k+1))) \longrightarrow \forall x\, P(x).$$

Showing that the first conjunct, $P(0)$, is true constitutes the **Basis step**, and showing that the second is true constitutes the **Inductive step**. To prove the inductive step, we shall assume that $P(k)$ is true for an arbitrary k and try to prove that $P(k+1)$ is true. The assumption, $P(k)$ is true is called the **Inductive Hypothesis (IH)**. Therefore, to prove $\forall x\, P(x)$ using mathematical induction, we shall do two things:

1. **Basis step:** Show that $P(0)$ is true.
2. **Inductive step:** Let k be an arbitrary element of $\mathbb{N}$, assume $P(k)$ is true and show that $P(k+1)$ is true.

Remark 5.1 *Sometimes we might wish to show that $P(x)$ is true for all $x \geq n_0$, with $n_0 > 0$. In that case, the basis step involves showing that $P(n_0)$ is true. Also, $k \in \mathbb{N}$ will be an arbitrary element with the property that $k \geq n_0$. we shall see examples of this case below.*

Here are two examples.

Example 5.2 Show
$$1 + 2 + \cdots + n = \frac{n(n+1)}{2},$$
for all $n \geq 1$, by mathematical induction.

Solution: *Note that here $P(n)$ is the entire equality above.*

Basis step: *We need to show that $P(1)$ is true. Note that the left-hand side gives 1 and the right-hand side is $\frac{1(1+1)}{2} = 1$. So, $P(1)$ is true.*

Inductive step:
Inductive hypothesis: $1 + 2 + \cdots + k = \frac{k(k+1)}{2}$
To show: $1 + 2 + \cdots + (k+1) = \frac{(k+1)(k+2)}{2}$

$$1 + 2 + \cdots + (k+1) = 1 + 2 + \cdots + k + (k+1)$$
$$= \frac{k(k+1)}{2} + (k+1), \text{ by IH}$$
$$= \frac{k(k+1) + 2(k+1)}{2}$$
$$= \frac{(k+1)(k+2)}{2}.$$

Example 5.3 Show

$$1^2 + 2^2 + \cdots + n^2 = \frac{n(n+1)(2n+1)}{6},$$

for all $n \geq 1$, by mathematical induction.

Solution: *Note that here $P(n)$ is the entire equality above.*

Basis step: *We need to show that $P(1)$ is true. Note that the left-hand side gives $1^2 = 1$ and the right-hand side is $\frac{1(1+1)(2.1+1)}{6} = 1$. So, $P(1)$ is true.*

Inductive step:
Inductive hypothesis: $1^2 + 2^2 + \cdots + k^2 = \frac{k(k+1)(2k+1)}{6}$
To show: $1^2 + 2^2 + \cdots + (k+1)^2 = \frac{(k+1)(k+2)(2k+3)}{6}$

$$1^2 + 2^2 + \cdots + (k+1)^2 = 1^2 + 2^2 + \cdots + k^2 + (k+1)^2$$
$$= \frac{k(k+1)(2k+1)}{6} + (k+1)^2, \text{ by IH}$$
$$= \frac{k(k+1)(2k+1) + 6(k+1)^2}{6}$$
$$= \frac{(k+1)(k(2k+1) + 6(k+1))}{6}$$
$$= \frac{(k+1)(2k^2 + 7k + 6)}{6}$$
$$= \frac{(k+1)(k+2)(2k+3)}{6}.$$

Here are a few more induction examples.

Example 5.4 Show

$$1 + 3 + 5 \cdots + (2n - 1) = n^2,$$

for all $n \geq 1$, by mathematical induction.

Solution: *Note that here $P(n)$ is the entire equality above.*

Basis step: *We need to show that $P(1)$ is true. Note that the left-hand side gives 1 and the right-hand side is $1^2 = 1$. So, $P(1)$ is true.*

Inductive step:
Inductive hypothesis: $1 + 3 + \cdots + (2k - 1) = k^2$
To show: $1 + 3 + \cdots + (2k - 1) + (2k + 1) = (k + 1)^2$

$$1 + 3 + \cdots + (2k - 1) + (2k + 1) = k^2 + (2k + 1), \text{ by IH}$$
$$= k^2 + 2k + 1$$
$$= (k + 1)^2.$$

Example 5.5 Show

$$1 + 2 + 2^2 + \cdots + 2^n = 2^{n+1} - 1,$$

for all $n \geq 0$, by mathematical induction.

Solution: *Note that here $P(n)$ is the entire equality above.*

Basis step: *We need to show that $P(0)$ is true. Note that the left-hand side gives 1 and the right-hand side is $2^{0+1} - 1 = 1$. So, $P(0)$ is true.*

Inductive step:
Inductive hypothesis: $1 + 2 + 2^2 + \cdots + 2^k = 2^{k+1} - 1$
To show: $1 + 2 + 2^2 + \cdots + 2^k + 2^{k+1} = 2^{k+2} - 1$

$$1 + 2 + 2^2 + \cdots + 2^k + 2^{k+1} = 2^{k+1} - 1 + 2^{k+1}, \text{ by IH}$$
$$= 2^{k+1} + 2^{k+1} - 1$$
$$= 2.2^{k+1} - 1$$
$$= 2^{k+2} - 1.$$

Example 5.6 Show

$$2^n < n!,$$

for all $n \geq 4$, by mathematical induction.

Solution: *Note that here $P(n)$ is the inequality above. Recall that*

$$n! = n.(n - 1).(n - 2).\ldots.3.2.1.$$

Basis step: *We need to show that $P(4)$ is true. Note that the left-hand side gives $2^4 = 16$ and the right-hand side is $4! = 24$. So, $P(4)$ is true.*

Inductive step:
Inductive hypothesis: $2^k < k!$
To show: $2^{k+1} < (k+1)!$

$$2^{k+1} = 2^k.2$$
$$< 2.k!, \quad by \ IH$$
$$< (k+1).k!, \quad as \ k \geq 4$$
$$= (k+1)!, \quad by \ definition \ of \ the \ factorial \ function.$$

Finally, we mention the very useful $\sum$ (sigma) notation which stands for summation. For example,

$$\sum_{i=1}^{4}(i+1) = (1+1) + (2+1) + (3+1) + (4+1).$$

Or,

$$\sum_{i=1}^{4} i^2 = 1^2 + 2^2 + 3^2 + 4^2.$$

Note that

$$\sum_{i=1}^{5} k = k + k + k + k + k = 5k.$$

Here are a few more induction examples:

Example 5.7 Show

$$\sum_{i=1}^{n}(i+1)2^i = n2^{n+1},$$

for all $n \geq 1$, by mathematical induction.

Solution: *Note that here $P(n)$ is the entire equality above.*

Basis step: *We need to show that $P(1)$ is true. Note that the left-hand side gives $(1+1)2^1 = 4$ and the right-hand side is $1.2^{1+1} = 4$. So, $P(1)$ is true.*

Inductive step:
Inductive hypothesis: $\sum_{i=1}^{k}(i+1)2^i = k2^{k+1}$

To show: $\sum_{i=1}^{k+1} (i+1)2^i = (k+1)2^{k+2}$

$$\sum_{i=1}^{k+1} (i+1)2^i = \sum_{i=1}^{k} (i+1)2^i + (k+2)2^{k+1}$$

$$= k.2^{k+1} + (k+2)2^{k+1} \ \ by \ IH$$

$$= k.2^{k+1} + k.2^{k+1} + 2.2^{k+1}$$

$$= 2.k.2^{k+1} + 2^{k+2}$$

$$= k.2^{k+2} + 2^{k+2}$$

$$= (k+1)2^{k+2}.$$

Definition 5.8 Let $a, b \in \mathbb{Z}$ and $a \neq 0$. We say that a *divides* b, denoted $a|b$, if there is a $k \in \mathbb{Z}$ such that $b = ka$, that is b is a *multiple* of a, or a is a *factor* of b.

For example $2|12, 3|12, 4|12$, but $5 \nmid 12, 12 \nmid 3$, etc.
Here are a few facts:

1. If $a|b$ and $a|c$, then $a|(b+c)$.
2. If $a|b$, then $a|bc$ for any $c \in \mathbb{Z}$.
3. If $a|b$ and $b|c$, then $a|c$.

It is very easy to prove these facts and we encourage you to use the definition of divisibility to do so.

Example 5.9 Show $3|(n^3 - n)$ for all $n \geq 1$, by mathematical induction.

Solution:

Basis step: *This is the case for* $n = 1$. *Note that* $3|0$ *and so we have verified the basis step.*

Inductive step:
Inductive hypothesis: $3|(k^3 - k)$
To Show: $3|(k+1)^3 - (k+1)$
Let us first simplify the expression $(k+1)^3 - (k+1)$.
$(k+1)^3 - (k+1) = k^3 + 3k^2 + 3k + 1 - k - 1 = (k^3 - k) + 3(k^2 + k)$.
Now $3|3(k^2 + k)$ *just because* $3(k^2 + k)$ *is a multiple of* 3 *and also by IH,* $3|(k^3 - k)$, *so using Fact 1, we see that* $3|(k+1)^3 - (k+1)$.

Exercises

Exercise 5.1 Prove by mathematical induction that

$$\frac{1}{1.2} + \frac{1}{2.3} + \cdots + \frac{1}{n(n+1)} = \frac{n}{n+1},$$

for all $n \geq 1$.

Exercise 5.2 Prove by mathematical induction that

$$3 + 3.5 + 3.5^2 + \cdots + 3.5^n = 3(5^{n+1} - 1)/4,$$

for all integers $n \geq 0$.

Exercise 5.3 Use mathematical induction to prove that

$$1^3 + 2^3 + \cdots + n^3 = [n(n+1)/2]^2,$$

for all $n \geq 1$.

Exercise 5.4 Prove by induction that $6^n + 4$ is divisible by 5 for every integer $n \geq 1$.

Chapter 6

Functions and Relations

We shall go back to functions and carry out a more in-depth study of them we shall also recall relations and discuss many useful properties of binary relations on a set A.

6.1 Functions

Recall that a function from a set A to a set B, denoted $f : A \longrightarrow B$ is a rule that to each element $x \in A$ assigns a unique element $f(x) \in B$. We call A the *domain* and B the *codomain* of the function f. Given $S \subseteq A$, $f(S) = \{f(x) \mid x \in S\}$ is called the *image of S under f*. Similarly, given $T \subseteq B$, $f^{-1}(T) = \{x \in A \mid f(x) \in T\}$ is called *the pre-image of T under f*. Note that $f^{-1}(T) \subseteq A$ for any $T \subseteq B$ and that $f(S) \subseteq B$ for any $S \subseteq A$. $f(A)$ is called the *range* of f, note that range of f is always a subset of its codomain.

Here is an example, let $A = \{a, b, c, d\}$ and $B = \{1, 2, 3, 4\}$ where

$$f(a) = 1, f(b) = 2, f(c) = f(d) = 3.$$

- The range of f is $f(A) = \{1, 2, 3\}$.
- $f(\{a, b\}) = \{1, 2\}$.
- $f^{-1}(\{4\}) = \emptyset$, while
- $f^{-1}(\{1, 4\}) = \{a\}$.

We say that a function $f : A \longrightarrow B$ is *one-to-one* if

$$x \neq y \longrightarrow f(x) \neq f(y),$$

117

for all $x, y \in A$. This means that f maps different elements to different images. Very often to prove a function is one-to-one we use the contrapositive of the formula above, that is:

$$f(x) = f(y) \longrightarrow x = y,$$

for all $x, y \in A$.

We say that $f : A \longrightarrow B$ is *onto*, if

$$\forall y \in B \, \exists x \in A \, f(x) = y.$$

This means that f covers every element in its codomain.

A function is said to be a *one-to-one correspondence* or a *bijection*, if it is both one-to-one and onto.

The example we had above is neither one-to-one (why?), nor onto (why?).

Example 6.1.1 Consider the function $f : \mathbb{N} \longrightarrow \mathbb{N}$ given by $f(x) = x^2$ where $\mathbb{N} = \{0, 1, 2, 3, \ldots\}$ is the set of natural numbers. Determine if this function is onto or one-to-one.

Solution: *This function is not onto, for example, if you pick $3 \in \mathbb{N}$, then there is no $x \in \mathbb{N}$ such that $f(x) = x^2 = 3$. This function however is one-to-one, we shall prove this as follows:*

Suppose $f(x) = f(y)$, then $x^2 = y^2$. Therefore, $\sqrt{x^2} = \sqrt{y^2}$ and thus $|x| = |y|$, but then $x = y$ as $x, y \geq 0$, This holds for any $x, y \in \mathbb{N}$ and hence the function is one-to-one. Recall that the absolute value *function $|.| : \mathbb{R} \longrightarrow \mathbb{R}, x \mapsto |x|$ is defined as*

$$|x| = \begin{cases} x, & \text{if } x \geq 0, \\ -x, & \text{if } x < 0. \end{cases}$$

Let us continue with a few more examples of functions.

Example 6.1.2 Consider the function $f : \mathbb{Z} \longrightarrow \mathbb{Z}$ given by $f(x) = x^2$ where $\mathbb{Z}$ is the set of integers. Determine if this function is onto or one-to-one.

Solution: *This function is not onto, for example, if you pick $3 \in \mathbb{Z}$, then there is no $x \in \mathbb{Z}$ such that $f(x) = x^2 = 3$. This function is not one-to-one either, for example, $f(2) = f(-2) = 4$ while $2 \neq -2$.*

Example 6.1.3 Consider the function $f : \mathbb{R} \longrightarrow \mathbb{R}$ given by $f(x) = x^2$ where $\mathbb{R}$ is the set of real numbers. Determine if this function is onto or one-to-one.

Solution: *This function is not onto, for example, if you pick $-3 \in \mathbb{R}$, then there is no $x \in \mathbb{R}$ such that $f(x) = x^2 = -3$. This function is not one-to-one either, for example, $f(2) = f(-2) = 4$ while $2 \neq -2$.*

Example 6.1.4 Consider the function $f : \mathbb{R}_0^+ \longrightarrow \mathbb{R}_0^+$ given by $f(x) = x^2$ where $\mathbb{R}_0^+$ is the set of non-negative real numbers. Determine if this function is onto or one-to-one.

Solution: *This function is onto: let $y \in \mathbb{R}_0^+$, define $x = \sqrt{y}$, note that x is well defined because $y \geq 0$. Now $x \geq 0$ too, and is a real number, so $x \in \mathbb{R}_0^+$, moreover $f(x) = f(\sqrt{y}) = (\sqrt{y})^2 = y$. This function is one-to-one too: $f(x) = f(y)$, so $x^2 = y^2$ and so $|x| = |y|$ but then $x = y$ as $x, y \in \mathbb{R}_0^+$.*

The point of these examples is to show that functions that have the same assignment rule but different domains and codomains can have very different properties. Thus, specification of a function consists of three parts: domain, codomain, and the assignment rule.

Given a set A, the *identity function on* A, denoted id_A is defined by $id_A(x) = x$, for all $x \in A$. Also given functions $f : A \longrightarrow B$ and $g : B \longrightarrow C$, we define the *composite* function, denoted $g \circ f : A \longrightarrow C$, by

$$(g \circ f)(x) = g(f(x)),$$

for all $x \in A$.

For example, let $f : \mathbb{R} \longrightarrow \mathbb{R}$ and $g : \mathbb{R} \longrightarrow \mathbb{R}$ be given as $f(x) = x^3$ and $g(x) = x^2 - 2x$, then $(g \circ f)(x) = g(f(x)) = g(x^3) = (x^3)^2 - 2(x^3) = x^6 - 2x^3$, whereas $(f \circ g)(x) = f(g(x)) = f(x^2 - 2x) = (x^2 - 2x)^3$. These two functions are clearly different, for example, $(g \circ f)(2) = 64 - 16 = 48$ while $(f \circ g)(2) = 0$.

Definition 6.1.5 A function $f : A \longrightarrow B$ is said to be *invertible* if there is a function $g : B \longrightarrow A$ such that

- $g \circ f = id_A$, and
- $f \circ g = id_B$.

The inverse of a function may not exist, for example, $f : \mathbb{R} \longrightarrow \mathbb{R}$ with $f(x) = x^2$ is not invertible. However, if f has an inverse, then the inverse will be unique and we denote it by f^{-1}.

The following fact will help to check if a function is invertible.

Theorem 6.1.6 *Let $f : A \longrightarrow B$ be a function, then f is invertible if and only if f is both onto and one-to-one, i.e. if it is a bijection.*

Example 6.1.7 Consider the function $f : \mathbb{R}_0^+ \longrightarrow \mathbb{R}_0^+$ given by $f(x) = x^2$. Is it invertible? If yes, find the inverse function.

Solution: *We know that f is a bijection, so it is invertible. To find the inverse:*

- *Write it in terms of y and x with y as a function of x: $y = x^2$,*
- *Solve for x: $x = \sqrt{y}$,*
- *Interchange x and y in step 2 above: $y = \sqrt{x}$. This is the inverse function.*

Check that $f^{-1}(x) = \sqrt{x}$ is the inverse of $f(x) = x^2$ with $f : \mathbb{R}_0^+ \longrightarrow \mathbb{R}_0^+$, using the definition of the inverse function above.

Example 6.1.8 Consider the function $f : \{a, b, c, d\} \longrightarrow \{1, 2, 3, 4\}$, with $f(a) = 4, f(b) = 2, f(c) = 3, f(d) = 1$. Is it invertible? If yes, find the inverse function.

Solution: *It can be easily seen that f is a bijection, so it is invertible. It can be easily checked that $f^{-1} : \{1, 2, 3, 4\} \longrightarrow \{a, b, c, d\}$ is given by $f^{-1}(1) = d, f^{-1}(2) = b, f^{-1}(3) = c, f^{-1}(4) = a$.*

Exercises

Exercise 6.1.1 Consider the function $f : \mathbb{N} \longrightarrow \mathbb{N}$ defined by $f(n) = n!$.

(i) Is f one-to-one?
(ii) Is f onto?
(iii) Let $E \subseteq \mathbb{N}$ be the set of all even natural numbers and $O \subseteq \mathbb{N}$ be the set of all odd natural numbers.

- Find $f(E) \cap O$.
- Find $f(O) \cap O$.

Exercise 6.1.2 Consider the function $s : \mathbb{N} \longrightarrow \mathbb{N}$ defined by $s(n) = n + 1$.

(i) Is f one-to-one?
(ii) Is f onto?
(iii) What is the range of f?

Exercise 6.1.3 Consider the function $g : \mathbb{R} \longrightarrow \mathbb{R}$ defined by $g(x) = x^3$. Is g invertible? If so, find its inverse.

Exercise 6.1.4 Consider the function $h : \{a, b, c\} \longrightarrow \{a, b, c\}$ defined by $h(a) = b, h(b) = a$ and $h(c) = c$. Is h invertible? If so, find its inverse.

6.2 Relations

Recall that a relation from a set A to a set B is a subset R of $A \times B$.

For example, let $A = \{a, b, c\}$ and $B = \{0, 1\}$, then $R_1 = \{(a, 0), (a, 1)\}$ is a relation, so are $R_2 = \emptyset$, and $R_3 = A \times B$. Note that every function is a relation, but the converse is not true, that is, not every relation will be a function. For example, with the sets A and B as above, R_1 is not a function.

A binary relation on a set A is just a relation from A to A, in other words a subset of $A \times A$.

We shall be interested in various properties of binary relations on a set A as follows.

Definition 6.2.1 A binary relation R on a set A is said to be

- *reflexive*, if for all $a \in A$, $(a, a) \in R$,
- *symmetric*, if for all $a, b \in A$, $(a, b) \in R \longrightarrow (b, a) \in R$,
- *antisymmetric*, if for all $a, b \in A$, $((a, b) \in R \wedge (b, a) \in R) \longrightarrow a = b$.

We can also write this definition contrapositively, then it will say that for all $a, b \in A$, $a \neq b \longrightarrow ((a, b) \notin R \vee (b, a) \notin R)$.

Let's consider a few examples.

Example 6.2.2 Consider the following relations:

1. $R_1 = \{(a, b), (b, c), (c, a)\}$, on the set $A = \{a, b, c\}$.

2. $R_2 = \{(x, y) \mid x, y \in \mathbb{Z}, (x = y) \vee (x = -y)\}$, on the set of integers $\mathbb{Z}$.

3. $(a, b) \in R_3$ iff a is b's brother, on the set of all people.

Determine whether each relation above is reflexive, symmetric, and/or antisymmetric.

Solution:

1. R_1 *is not reflexive as e.g.,* $(a, a) \notin R_1$. *It is not symmetric because although* $(a, b) \in R_1$, $(b, a) \notin R_1$. *But* R_1 *is antisymmetric, to show this we check that in all three cases where a pair is in* R_1, *the statement defining antisymmetry holds true.*

2. R_2 *is reflexive, because for all* $x \in \mathbb{Z}$, $(x, x) \in R_2$. *It is also symmetric: suppose* $(x, y) \in R_2$, *then* $x = y$ *or* $x = -y$, *so* $y = x$ *or* $y = -x$, *thus* $(y, x) \in R_2$. *However,* R_2 *is not antisymmetric because* $(1, -1) \in R_2$ *and* $(-1, 1) \in R_2$, *but* $1 \neq -1$.

3. R_3 *is not reflexive, as nobody is his own brother. It is not symmetric, suppose* $(John, Alice) \in R_3$, *but* $(Alice, John) \notin R_3$. *It is also not antisymmetric, suppose* $(John, Jim) \in R_3$, *then surely* $(Jim, John) \in R_3$, *but Jim and John are not the same people.*

We can represent relations on finite sets as graphs. Suppose R is a relation on a finite set A. Then our graph will have vertices labelled with the elements of the set A and there will be an edge from a vertex a to another vertex b iff $(a, b) \in R$. A loop is a cycle of length 1. A 2-cycle is a cycle of length 2, that is it consists of two distinct edges. We shall call 2-cycles, just cycles. In terms of graphs:

- Reflexivity means that all vertices have loops.
- Symmetry means that every edge is either part of a cycle or is a loop.
- Antisymmetry means that there are no cycles.

Apply these geometric definition and characterization to R_1 in the example above.

Note that symmetry and antisymmetry are not opposite notions. We can have relations that are both symmetric and antisymmetric, for example, $\{(a, a), (b, b)\}$ on the set $A = \{a, b, c\}$. Also we can have relations that are neither symmetric nor antisymmetric, for example, $S = \{(a, b), (b, a), (a, c)\}$ on the set A.

Let's consider a few more examples.

Example 6.2.3 Consider the relation D_1 on the set of non-zero integers, $\mathbb{Z} - \{0\}$ defined by, $(a, b) \in D_1$ iff $a|b$. Determine whether D_1 is reflexive, symmetric, and/or antisymmetric.

Solution: D_1 *is reflexive as* $a = 1.a$ *for all* $a \in \mathbb{Z} - \{0\}$. *However,* D_1 *is not symmetric, for example,* $2 \mid 12$ *but* $12 \nmid 2$. D_1 *is not antisymmetric because, for example,* $(1, -1) \in D_1$ *and* $(-1, 1) \in D_1$ *but* $1 \neq -1$.

Example 6.2.4 Consider the relation D_2 on the set of integers, $\mathbb{Z}$, defined by $(a, b) \in D_2$ iff $a|b$. Determine whether D_2 is reflexive, symmetric, and/or antisymmetric.

Solution: D_2 *is reflexive. It is not symmetric, for example,* $2 \mid 12$ *but* $12 \nmid 2$. D_2 *is not antisymmetric because, for example,* $(1, -1) \in D_2$ *and* $(-1, 1) \in D_2$ *but* $1 \neq -1$.

Example 6.2.5 Consider the relation D_3 on the set of positive integers, $\mathbb{Z}^+$, defined by, $(a, b) \in D_2$ iff $a|b$. Determine whether D_3 is reflexive, symmetric, and/or antisymmetric.

Solution: D_3 *is reflexive as* $a = 1.a$ *for all* $a \in \mathbb{Z}^+$. *However,* D_3 *is not symmetric, for example,* $2 \mid 12$ *but* $12 \nmid 2$. D_3 *is antisymmetric: suppose* $(a, b) \in D_3$, *then* $a \mid b$ *and so* $b = k.a$ *for some positive integer* k. *Also suppose* $(b, a) \in D_3$, *so* $b \mid a$ *and thus we have* $a = k'.b$. *From these equations we get* $a = k'.b = k'.k.a$ *and so* $k'.k = 1$, *the only solution for this equation over* $\mathbb{Z}^+$ *is* $k = k' = 1$ *and so* $a = b$.

The point of these examples is that, although in each case the relations are defined using the same rule (that is, divisibility), they have different properties because they are defined on different domains.

Definition 6.2.6 We say that a relation R on a set A is *transitive* if for all $a, b, c \in A$,

$$((a, b) \in R \wedge (b, c) \in R) \longrightarrow (a, c) \in R.$$

For example, the relation $R = \{(1, 2), (2, 3), (2, 2), (3, 1)\}$ on the set $A = \{1, 2, 3\}$ is not transitive, because although $(1, 2)$ and $(2, 3)$

are both in R, $(1,3) \notin R$. On the other hand, the relation

$$S = \{(1,2), (1,3), (2,3), (3,2), (1,1), (2,2), (2,1), (3,3), (3,1)\}$$

is transitive.

Example 6.2.7 Consider the relation D_3 above on $\mathbb{Z}^+$. Is D_3 transitive?

Solution: *Suppose $(a,b) \in D_3$, then there is a k such that $b = k.a$ and suppose that $(b,c) \in D_3$, then there is k' such that $c = k'.b$. So, we have $c = k'.b = k'.k.a$, so c is a multiple of a and therefore $(a,c) \in D_3$.*

Example 6.2.8 Let's look at an old example: Is

$$R = \{(x,y) \mid x,y \in \mathbb{Z}, (x = y) \vee (x = -y)\}$$

transitive?

Solution: *Suppose $(x,y) \in R$, then $x = y$ or $x = -y$, also suppose $(y,z) \in R$, then $y = z$ or $y = -z$. From these four equations we get that $x = z$ or $x = -z$ and thus $(x,z) \in R$. And so R is transitive.*

Definition 6.2.9 Given the relations $R : A \longrightarrow B$ and $S : B \longrightarrow C$, we define their *composite*, denoted $S \circ R : A \longrightarrow C$ by

$$(a,c) \in S \circ R \text{ iff } \exists b \in B \, ((a,b) \in R \wedge (b,c) \in S).$$

In other words, this says that a and c are related by $S \circ R$, if there is a mediating b such that a and b are related by R, and b and c are related by S.

Here is an example. Suppose $A = \{1,2,3\}, B = \{a,b,c\}, C = \{x,y,z\}$ and that $R = \{(1,a), (2,a), (2,b)\}$ and $S = \{(a,x), (b,y), (b,z), (c,z)\}$. Then $S \circ R = \{(1,x), (2,x), (2,y), (2,z)\}$.

Relations arise in all areas of mathematics both in pure and in applied mathematics. You will see many instances of phenomena modeled by relations, and thus a working knowledge of properties of relations will help you analyze the structures you will be dealing with.

Exercises

Exercise 6.2.1 Let $A = \{a, b, c, d\}$.

- Find a non-empty binary relation on A that is *both* symmetric and antisymmetric and call it R.
- Find $R \circ R$. Is $R \circ R$ symmetric?

Exercise 6.2.2 Let $A = \{0, 1, 2, 3, 4\}$.

- Find a non-empty binary relation on A that is *neither* symmetric *nor* antisymmetric.
- Can you find a binary relation on A that is *both* asymmetric (we say that a relation R on a set A is *asymmetric*, if for all $a, b \in A$, $(a, b) \in R \longrightarrow (b, a) \notin R$) and not antisymmetric?

Exercise 6.2.3 Let $A = \{a, b, c, d\}$.

- Find a non-empty symmetric relation on A and call it R. Find a non-empty antisymmetric relation on A and call it S.
- Is $R \cap S$ symmetric?
- Is $R \cup S$ symmetric?

Exercise 6.2.4 Let $A = \{1, 2, 3, 4\}$ and $B = \{a, b, c, \}$.

- Find a non-empty antisymmetric relation on $A \times B$.
- Find a non-empty symmetric relation on $A \times B$.
- Find a transitive relation on $A \times B$.

Chapter 7

Graph Theory and Its Applications

Graphs are one of the most interesting and useful mathematical objects. They are interesting for mathematicians for the vast number of places where they show up and the same is true for practitioners in many other fields where graphs are used for modeling. In this chapter, we will introduce graphs using relations that we saw in the previous chapter. There are many books on graph theory which you can read for your particular needs. Here we shall try to give a brief introduction to graph theory.

We shall start with some basic definitions first.

7.1 Basic Definitions

Recall the definition of a relation from the previous chapter. A *directed graph* G consists of a pair of sets (V, E) where V is a non-empty set called the *vertices or nodes* of the graph G, and E is a binary relation on V called the set of *edges* of the graph G. Given a directed graph $G = (V, E)$, the *order* of G is defined to be the cardinality of its vertex set, that is $|V|$. The cardinality of the set of edges, $|E|$ gives you the number of edges in the graph. Note that $|E| \leq |V| \times |V|$. As $V \times V$ is the largest binary relation on V.

For example, $G = (\{1, 2, 3\}, \{(1, 2), (2, 3), (3, 1)\})$ is a graph with three vertices and three edges. It is often very convenient to use a pictorial representations of graphs. In doing so we draw circles for vertices and arrows between them for the edges. So, for example, the graph G above can be pictorially represented as

This graph has three vertices (nodes) and three edges. It should be clear that we can go back from a given drawing of a graph to its definition in terms of sets.

For example,

defines the graph $K = (\{1, 2, 3\}, \{(2, 1), (1, 3), (2, 3)\})$.

As you can see from the definition of the directed graph, the edges have directions associated with them. That is why we call these *directed* graphs. Such structures can be used to model applications where the relationship between entities are directional and may not always be reciprocated. For example, you could model lending money by a person A to a person B by an edge from A to B, etc. We may have loops in a graph, that is, an edge that connects a node to itself. For example, the graph $G_1 = (\{1, 2, 3\}, \{(1, 2), (2, 3), (1, 1), (3, 3)\})$ below has two loops.

Moreover, sometimes we may have multiple edges between vertices. In order to represent them we will have to augment our model to include multiplicities for each member of the set of edges. We shall not pursue this direction. A graph is *simple* if there are no self-loops, and no multiple edges. Given an edge (i, j) in a graph G we say that node i is *adjacent* to node j. Note that if there is no edge from j to i, then j is not adjacent to i. In other words, because of directionality of the edges adjacency is not symmetric. So, in the graph G above, 1 is adjacent to 2, but 2 is not adjacent to 1. We say that 2 is an *out-neighbor* of 1, and 3 is an *in-neighbor* of 1. A formal definition can be easily given but the notion of neighborhood is not very often used in directed graphs. We will give a formal definition when we study undirected graphs.

Given a directed graph $G = (V, E)$ and a vertex $v \in V$, the *out-degree* of v, denoted $deg^+(v) = |\{w \in V \,|\, (v, w) \in E\}|$. That is, the number of vertices w such that there is a directed edge from v to w. Similarly, the *in-degree* of v, denoted $deg^-(v) = |\{w \in V \,|\, (w, v) \in E\}|$. That is, the number of vertices w such that there is a directed edge from w to v. For example, in the graph G_1 above, $deg^+(2) = 1$ because there is only one vertex (vertex 3) with an edge from 2 to it. Also, $deg^-(3) = 2$ because there are two vertices with edges into vertex 3.

There is a very useful and important result in graph theory that we mention here without proof. It is called the *handshaking theorem* which for directed graphs has the following form.

Theorem 7.1.1 *Let $G = (V, E)$ be a directed graph, then*

$$\sum_{v \in V} deg^+(v) = \sum_{v \in V} deg^-(v) = |E|.$$

Not all graphs are directed. In order to model undirected graphs, we define the set of edges E as a subset of $\mathcal{P}_2(V)$, the set of all non-empty subsets of V of cardinality at most 2. Here is an example; $G_2 = (\{1, 2, 3\}, \{\{1, 2\}, \{2, 3\}\})$ with three vertices and two edges. We can represent G_2 pictorially by

Let $G = (V, E)$ be an undirected graph. Given an edge $\{i, j\} \in E$, we say that i and j are adjacent. The neighbors of a vertex $v \in V$, denoted $N_G(v)$ is defined to be the set of all vertices adjacent to v, mathematically

$$N_G(v) = \{w \in V \,|\, \{v, w\} \in E\}.$$

The degree of a vertex v, denoted $deg(v)$ is the number of incident edges on $v \in V$, formally

$$deg(v) = |\{w \,|\, \{v, w\} \in E\}|.$$

The Handshaking Theorem above takes the following form for undirected graphs.

Theorem 7.1.2 *Let $G = (V, E)$ be an undirected graph, then*

$$\sum_{v \in V} deg(v) = 2|E|.$$

Example 7.1.3 How many edges are there in an undirected graph with 10 vertices each of degree 6?

Solution: *The sum of the degrees of the vertices is $6.10 = 60$. According to the Handshaking Theorem $2|E| = 60$, so there are 30 edges.*

An undirected graph $G = (V, E)$ is said to be a *complete graph* if $\{i, j\} \in E$ for all $i, j \in V, i \neq j$.

Example 7.1.4 Consider the graph G below. Find the degrees of all the vertices, and $N_G(2)$.

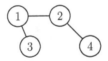

Solution:

$$deg(1) = deg(2) = 2, deg(3) = deg(4) = 1, \quad and \quad N_G(2) = \{1, 4\}.$$

A *subgraph* H of $G = (V, E)$ is a graph $H = (W, F)$ where $W \subseteq V$ and $F \subseteq E$. The *induced subgraph of G* formed by the set of vertices $W \subseteq V$ is a subgraph where we take all the edges in E that are adjacent on any of the vertices in W.

Example 7.1.5 Given the graph G below, list three subgraphs of G.

Solution: *Here are three subgraphs of G.*

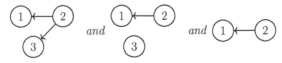

Example 7.1.6 Consider the graph G below.

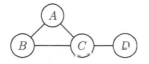

Determine if each of the following graphs G_1, G_2, G_3, and G_4 is a subgraph of G.

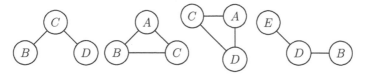

Solution:

- G_1 *is a subgraph of G because* $\{B, C, D\} \subseteq \{A, B, C, D\}$ *and all edges of G_1 are edges in G.*
- G_2 *a subgraph of G because* $\{A, B, C\} \subseteq \{A, B, C, D\}$ *and all edges of G_2 are edges in G.*
- G_3 *is not a subgraph of G because* (A, D) *is not an edge in G.*
- G_4 *is not a subgraph of G because* $\{E, B, D\} \nsubseteq \{A, B, C, D\}$.

Example 7.1.7 Determine if each of the following graphs is directed or undirected? Do self-loops or multiple edges make sense in each graph?

a. LinkedIn network.
b. The X network.
c. TikTok network.

Solution:

a. *LinkedIn, a professional social network and career development platform that enables members to build their professional network. This network is directed, signifying that when Member A follows Member B, there is a direction from A to B. Self-loops and multiple edges are irrelevant in LinkedIn because self-following and duplicate connections do not apply.*
b. *The X social network, formerly known as Twitter, enables its members to follow other members. The X network is directed because Member A can follow Member B, and Member B, in turn,*

may follow Member C or A. Self-loops and multiple edges are not applicable within the X network because nobody follows themselves or the same member multiple times.

c. *TikTok is a popular social media platform known for its short-form video content and user-generated creativity. TikTok network functions as a directed social network. Users can follow other creators, establishing clear one-way relationships. Self-loops and multiple edges are not relevant within TikTok's context because users do not follow themselves or others more than once.*

Example 7.1.8 Given a Facebook Friends graph below, answer the following questions.

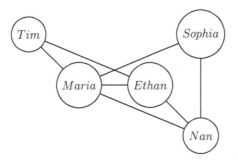

a. Who are Ethan's neighbors?
b. Is Sophia Ethan's neighbor?
c. What is the degree of Ethan?

Solution:

a. *The neighbors of Ethan are $\{Maria, Nan, Tim\}$.*
b. *No, Sophia is not Ethan's neighbor.*
c. *The degree of Ethan is 3.*

Example 7.1.9 Given a Facebook Friends graph as in the previous example, answer the following questions.

a. What is the number of edges?
b. What is the degree of each vertex?
c. What is the sum of degrees of all vertices? Is it equal to twice the number of edges?

Solution:

a. *The number of edges is 7.*
b. *deg(Maria) = 4, deg(Tim) = 2, deg(Sophia) = 2, deg(Nan) = 3, and deg(Ethan) = 3.*
c. *The total number of degrees of all vertices is 4 + 2 + 2 + 3 + 3 = 14. It is equal to twice the number of edges, 7.*

We close this section by introducing the notion of connectivity in a graph. Consider an undirected graph G, we say that G is *connected* if for every pair of vertices $v, w \in V$, there is a path (i.e., sequence of edges) from v to w. This is a bit more complicated to define for a directed graph. We shall only give one definition of connectivity which in some books is referred to as strong connectivity. We say that a directed graph $G = (V, E)$ is *connected* if for any pair $v, w \in V$ there is a directed path from v to w and a directed path from w to v. Let us look at some examples of each case.

Example 7.1.10 Consider the graph G below. Is G connected?

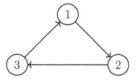

Solution: *G is connected because no matter which pair of distinct nodes you pick there is always a path from one to the other and vice-versa. For example, pick vertices 1 and 3, then we can go (1, 2), (2, 3) to get from 1 to 3, and we can go (3, 1) to get from 3 to 1.*

Example 7.1.11 Consider the graph G below. Is G connected?

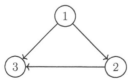

Solution: *G is not connected, for example, if you pick 1, 3, then we cannot go from 3 to 1.*

Example 7.1.12 Consider the graph G below. Is G connected?

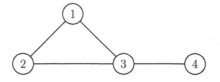

Solution: *G is connected because no matter which pair of distinct nodes you pick there is always a path from one to the other. For example, pick 1, 4, then we can go $\{1,3\}, \{3,4\}$ to get from 1 to 4.*

Example 7.1.13 Consider the graph G below. Is G connected?

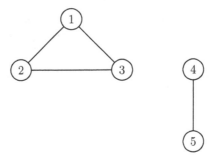

Solution: *G is not connected because you cannot get to node 4 or 5 starting from, say, node 1.*

Example 7.1.14 Determine if each of the graphs G_1 and G_2 below is connected.

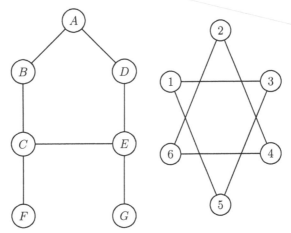

Solution: G_1 *is connected because there is a path between every pair of vertices.* G_2 *is not connected. Odd-numbered nodes (1, 3, and 5) have edges only to odd-numbered nodes and even-numbered nodes (2, 4, and 6) have only edges to even-numbered nodes. For example, there is no path between node 1 and node 4; there is no path between node 2 and node 5.*

Exercises

Exercise 7.1.1 Given the graph G below, answer the following questions:

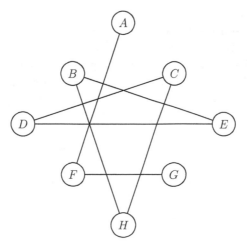

a. Find a path from D to B.
b. Is the graph connected?

Exercise 7.1.2 Given the graph below, answer the following questions:

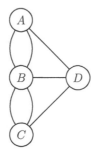

a. What is the neighborhood of node A and node B?
b. What is the degree of each vertex?
c. What is number of edges?
d. Is the sum of degrees of all vertices twice the number of edges?

7.2   Graph Isomorphism

As you have seen before in this book, once we introduce a new mathematical concept we would like to discuss the sameness question. So, what is the right notion of graph sameness? As it turns out and makes intuitive sense, we shall call two graphs the same if they have the same number of nodes and edges with the same structure of connectivity. So, essentially two graphs are the same if they only differ in the names (labels) of the nodes and edges. Here is how we shall make our definition mathematical. Suppose we are given two directed graphs $G = (V, E)$ and $H = (W, F)$, we say that G and H are isomorphic, denoted $G \cong H$ if there is a bijective map $f : V \longrightarrow W$ such that for every two vertices $u, v \in V$,

$$(u, v) \in E \text{ iff } (f(u), f(v)) \in F.$$

As for undirected graphs, suppose we are given two undirected graphs $G = (V, E)$ and $H = (W, F)$, we say that G and H are isomorphic, denoted $G \cong H$ if there is a bijective map $f : V \longrightarrow W$ such that for every two vertices $u, v \in V$,

$$\{u, v\} \in E \text{ iff } \{f(u), f(v)\} \in F.$$

So, if you are asked to check if two graphs are isomorphic, you either find a bijection respecting the connectivity and answer in affirmative or find a counterexample to two graphs being isomorphic.

Let us look at some examples.

Example 7.2.1 Are the following two graphs G and H isomorphic?

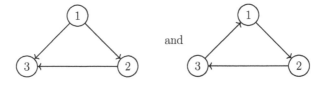

Solution: *These two graphs have the same number of nodes and edges but the out-degree of node 1 in G is 2 but there are no nodes in H with out-degree 2. So, you cannot possibly find a function from G to H that can preserve the connectivity pattern. Therefore, G and H are not isomorphic.*

Now let us consider the following undirected graphs.

Example 7.2.2 Are the following two graphs G and H isomorphic?

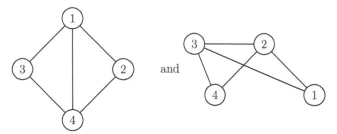

and

Solution: *These two graphs have the same number of nodes and edges. Consider the function $f : \{1, 2, 3, 4\} \longrightarrow \{1, 2, 3, 4\}$ given by $f(1) = 2, f(2) = 4, f(3) = 1,$ and $f(4) = 3.$ You can check that this map is a bijection and preserves the connectivity structure of the graphs. Therefore, G and H are isomorphic. This map is not the only one that proves that graphs are isomorphic. Try to find a different one.*

Example 7.2.3 Are the following two graphs G and H isomorphic?

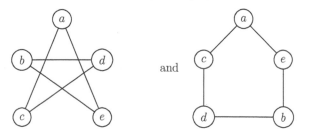

and

Solution: *Consider the map $f : \{a, b, c, d, e\} \longrightarrow \{a, b, c, d, f\}$ defined as the identity map, that is, it maps every letter to itself. Then, you can easily check that this map is a bijection and preserves the connectivity structure of these graphs. Therefore, G ad H are isomorphic.*

Exercises

Exercise 7.2.1 Which pairs of the following graphs G, H, K are isomorphic?

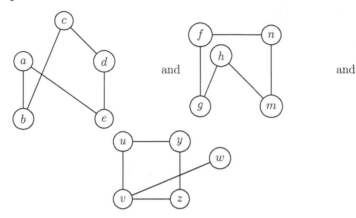

and   and

Exercise 7.2.2 Given the graph H below, find three of its subgraphs.

7.3   Paths in a Graph

Quite often in graph theory we are interested in connectivity questions. For example, if you consider the graph of cities and roads between them, connectivity becomes a very important issue. We briefly looked at the notion of a connected graph in Section 7.1. In this section, we take on a more detailed study of graph connectivity and related results. Given a graph G, a path in G is a sequence of edges that connects vertices. Of course in order for a path to be traversed in a directed graph, the orientation should be properly connecting the vertices on the graph. For example, consider the graph G below.

Here $(1, 2), (2, 3)$ is a path but $(3, 2)$ is not a path simply because there are no edges going from vertex 3 to vertex 2. On the other hand, in the undirected graph H below.

$\{1, 2\}, \{2, 3\}; \{2, 3\};$ and $\{3, 2\}$ are all paths of different lengths. A *circuit* is a path that starts and ends in the same vertex. For example, in the graph G above $(1, 2), (2, 3), (3, 1)$ is a circuit. A circuit or path is *simple* if it does not contain the same edge more than once.

Example 7.3.1 In the graph below give examples of simple paths, non-simple paths, circuits and non-simple circuits.

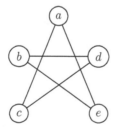

Solution:

- *Simple path:* $\{a, c\}, \{c, d\}, \{d, b\}$.
- *Simple circuit:* $\{a, c\}, \{c, d\}, \{d, b\}, \{b, e\}, \{e, a\}$.
- *Non-simple path:* $\{a, c\}, \{c, d\}, \{d, c\}$.
- *Non-simple circuit:* $\{a, c\}, \{c, d\}, \{d, b\}, \{b, d\}, \{d, b\}, \{b, e\}, \{e, a\}$.

Example 7.3.2 Find some simple and non-simple circuits in G below.

Solution:

- *Simple circuit*: $\{a, c\}, \{c, d\}, \{d, b\}, \{b, a\}$.
- *Non-simple circuit*: $\{a, c\}, \{c, d\}, \{d, b\}, \{b, c\}, \{c, a\}$.

The Swiss mathematician and the most prolific mathematician of the 17th century, Leonhard Euler famously solved the Königsberg bridges problem using a graph theoretic model. You must have heard about this problem but just in case, we recall the problem here. Königsberg had 7 bridges over the Pregolya river (google it to see for yourself) that connected various islands to each other.

People of Königsberg took long walks in the area and wondered if it were possible to start at any location, traverse all the bridges only once and get back to the start location. Euler modeled this as a graph shown below.

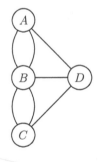

Note that here A, B, C and D represent the land and the edges represent the bridges. So, there are 4 pieces of land connected by 7 bridges. So, let us think about this. How does this modeling help solve the problem? How can we translate our original problem into this graph setting? A little bit of thought shows that traversing all the bridges (edges of the graph) and getting back to your starting point means that we are looking for a circuit. But we want to traverse each bridge only once, so we are looking for a simple circuit. Moreover, we want to traverse all 7 bridges so we are looking for a special circuit that goes over each and every edge of a graph. This is a special kind of a circuit called an *Euler circuit*. Formally,

Definition 7.3.3 An *Euler circuit* in a graph G is a simple circuit that contains all the edges of the graph G.

So, now the problem is translated to the following: Does the graph of the seven bridges have an Euler circuit?

Let us look at some examples first.

Example 7.3.4 Does the graph G below have an Euler circuit?

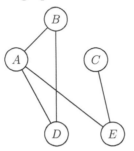

Solution: *No matter how we try, we cannot find a way to traverse all the edges and get back to our starting point.*

Example 7.3.5 Does the graph H below have an Euler circuit?

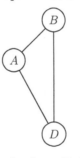

Solution: *Clearly this graph has Euler circuits, for example,* $\{B, A\}, \{A, D\}, \{D, B\}$.

But how are we going to know if there is an Euler circuit when a graph is much larger and we cannot use trial and error method to find such circuits or their lack thereof? It turns out that there is a very useful theorem that supplies the answer.

Theorem 7.3.6 *A connected multigraph with at least two vertices has an Euler circuit if and only if all vertices have even degrees.*

So, now you can look at the graph of the seven bridges and see that there are no Euler circuits because there is at least one node of odd degree.

Exercises

Exercise 7.3.1 In the following graph, does an Euler circuit exist for node A? Find an Euler circuit if it exists.

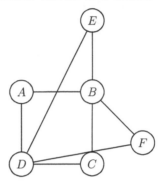

Exercise 7.3.2 Give some examples of simple and non-simple circuits in H.

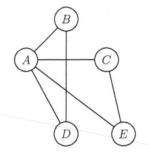

Appendix A

Translation to Propositional Logic

Here are some common stylistic variants of the five logical connectives we study:

- It is not the case that A: translated $\neg A$

  It is false that A,
  not A,
  A is false,
  A is not true,
  A does not happen,
  A fails.

- A and B: translated $(A \wedge B)$

  Both A and B,
  A but B,
  A although B,
  A, however B,
  A, whereas B,
  A, and also B,
  A besides B,
  A, nevertheless B,
  A, nonetheless B,
  A, even though B,
  not only A but also B,
  A in spite of the fact that B,
  A in as much as B,
  A, while B,
  A, since B,

     A, as B,
     A together with B,
     A as well as B,
     the conjunction of A and B.

- A or B: translated $(A \lor B)$

     either A or B,
     A or else B,
     A or, alternatively B,
     A otherwise B,
     A with the alternative that B,
     A unless B.

- If A, then B: translated $(A \longrightarrow B)$

     if A, B,
     B, if A,
     given that A, (it follows that) B,
     in case A, B,
     insofar as A, B,
     A leads to B,
     whenever A, B,
     A only if B,
     only if B, A,
     provided that A, B,
     B, provided that A,
     so long as A, B,
     A is a sufficient condition for B,
     B is a necessary condition for A.

- A if and only if B: translated $(A \leftrightarrow B)$

     A exactly in case B,
     A just in case B,
     A when and only when B,
     A is equivalent to B,
     A is a necessary and sufficient condition for B.

Appendix B

Translation to Predicate Logic

Here are some common stylistic variants for translations involving universal and existential quantifiers:

English sentence	Translation
Everything is A Everything is an A	$\forall x A(x)$
Something is an A There is an A At least one A exists There are A's	$\exists x A(x)$
Nothing is A A's do not exist	$\neg(\exists x A(x))$ or $\forall x(\neg(A(x)))$
Something is not A There is a non-A.	$\exists x(\neg A(x))$
Every A is a B Every A is B Each A is a B All A's are B's A's are B's Any A is a B	$\forall x(A(x) \longrightarrow B(x))$
No A is a B No A's are B's None of the A's are B's	$\forall x(A(x) \longrightarrow \neg B(x))$ or $\neg(\exists x(A(x) \wedge B(x)))$

(Continued)

(Continued)

English sentence	Translation
Some A is a B At least one A is a B There exists an A which is a B	$\exists x(A(x) \wedge B(x))$
Some A is not a B Some A's are not B's Some A's are not B	$\exists x(A(x) \wedge \neg B(x))$

Index